MEXICO'S
AGRICULTURAL
DILEMMA

MEXICO'S AGRICULTURAL DILEMMA

P. Lamartine Yates

THE UNIVERSITY OF ARIZONA PRESS
TUCSON, ARIZONA

About the author ...

PAUL LAMARTINE YATES became adviser on economic and social matters to the Banco Nacional de México in 1974 and to the Statistical
Office of the Mexican government in 1978. He began his association
with the agricultural and economic communities of Mexico as an
adviser to the Mexican government in 1957. A 1930 graduate of St.
John's College, Cambridge, he was Regional Director for Europe for
the Food and Agriculture Organization from 1961 to 1970 and has
served in various capacities for government and civilian organizations in Mexico, the Carribean, Europe, and the Middle East. His
publications include *Forty Years of Foreign Trade, El Desarrollo
Regional de México,* and *El Campo Mexicano,* as well as other books
and articles about world agriculture and economics.

THE UNIVERSITY OF ARIZONA PRESS

This book was set in 10/11 V.I.P. Times Roman

Library of Congress Cataloging in Publication Data

Lamartine Yates, Paul.
    Mexico's agricultural dilemma.

    Abridged, rev. and updated version of the au-
thors El Campo mexicano.
    Bibliography: p.
    Includes index.
    1. Agriculture—Economic aspects—Mexico.
2. Agriculture—Mexico. 3. Agriculture—Social
aspects—Mexico. I. Title.
HD1792.L35213   338.1'0972   81-10279

ISBN 0-8165-0734-1   AACR2
ISBN 0-8165-0733-3 (pbk.)

to
Sheila

Contents

TABLES

ILLUSTRATIONS

Foreword

Paul Yates' *Mexico's Agricultural Dilemma* could not have arrived on the world scene at a more propitious time. Mexico, indeed, presents us with a dilemma in the form of a question: how can a country's agricultural sector resolve the classical problem of providing food and fiber for a burgeoning population—or, increased self-sufficiency—and, at the same time, increase real productivity within the sector, provide for more employment and, simultaneously, generate foreign exchange under conditions of international comparative advantage? No easy task, to which the principal text will attest. The recent oil boom in Mexico adds a special dimension. Yates does not dodge the difficult issues, as for example in Chapter 8 where he boldly proposes a "new agrarian reform" for the festering sores of the *ejido* and land proprietorship, which have bogged down and destabilized Mexican agricultural production for decades.

Other subject matter is attacked with authority, and head on. Some will not agree with Yates' implied solutions, but hardly any can disagree with his clarity of statement. He brings to this work the simple declarative sentence structure and analytical language for which he became well known in earlier works, such as *Forty Years of Foreign Trade*. His treatment of agricultural prices, credit expansion, basic crops, and extension services—along with collectives—are among many others which are analyzed with policy reorientation in mind. The author argues that a major internal agricultural crisis, having been building since the mid-1960s, demands such policy redirection.

The Mexican agricultural dilemma has more universal implications than Yates would lead us to believe with his emphasis on the apparent uniqueness of Mexican social structure, agricultural idiosyncrasies, physical and human programs, and the general diversity of circumstances in that country. Couldn't it be said of many developing countries that their case is "different"? What I observe and conclude, without getting technical or disputatious, is that this book has broader application for the agriculture of developing countries than is implied by this "case study."

Mexico's successes as well as her obstacles to increased productivity are instructive to many countries. Certainly the institutional structures and norms of Mexico bear careful inquiry by those who would try to improve both efficiency and equity within the agricultural sectors as well as between agriculture and the rest of the economy in the third world. In short, Yates tells us more in this book than he suggests in the general introduction, the first chapter.

This study also has important implications for Mexico–United States relations. Our two countries are bound by a long historic relationship, most of which has been friendly, a seventeen-hundred-mile border, and mutually interdependent economies. The agriculture of Mexico is both complementary and competitive with that of the United States. The United States exports to Mexico considerable corn and other feed grains, wheat, hides, dried beans, soybeans, and soybean products, and minor amounts of other products. It imports from Mexico, in addition to large amounts of fresh fruits and vegetables, coffee, live cattle (feeder type), and beef. Traditionally, the United States has had an unfavorable balance of agricultural trade with Mexico, but during several recent years that balance has improved relatively for a variety of reasons. Agricultural producers in both countries have a mutual interest in the markets in the other.

Clearly, unemployment and agricultural productivity are two of Mexico's problems with which the United States should be concerned. It is also clear that U.S. interests in a healthy Mexican agricultural economy extend far beyond the relatively cheap, stable, and wholesome supply of tomatoes U.S. consumers receive during winter months from south of the border. There are, in addition, the ultimate and sufficient conditions in international trade: reciprocity of demand. In short, in order to buy, one must sell. For example, the millions of dollars earned by the Mexican fresh vegetable industry come back to the United States in hundreds, if not thousands, of direct and indirect ways via the trade process.

Finally, it is fitting that this study is available in English to readers and researchers on the subject of Mexican agriculture. Its notable, two-volume predecessor in Spanish, Yates' *El Campo Mexicano*, published in 1978, has had wide circulation in international circles, particularly in South America. With this current volume, therefore, a wider audience will come to know, not only of Mexico's dilemma in *el campo*, but will be able to assess the significance of that dilemma for other countries, for other agricultural sectors, for private firms, and even for personal decisions.

JIMMYE S. HILLMAN
Chairman
Department of Agricultural Economics
University of Arizona

Preface

In the summer of 1978 this author published in Spanish *El Campo Mexicano*, a two-volume survey of Mexican agriculture. That study resulted from nearly three years of research in the literature and in the field, including visits to all parts of the Mexican Republic and talks with many hundreds of farmers and government officials. In the difficult task of drawing conclusions from the information obtained, the author benefited incalculably from the guidance and wisdom of Ing. Emilio Alanís Patiño, who generously shared his profound experience of Mexico's farm sector and its history and who also made available his richly stocked library. To him a great debt of gratitude is owed. Ing. Miguel Sánchez Fuentes provided valuable assistance in making field contacts, and in collecting and interpreting statistics Lic. Miguel-Angel Castro Pedraza was equally helpful. Thanks are also due to the University of Arizona Press for effective publication.

The present volume is an abridged, revised, and updated version of the original work. It includes everything from the former study likely to be of interest and use to the English-speaking reader, and the opportunity has been taken of inserting the most recent data. Special consideration has been given to the changes which petroleum is likely to impose on the Mexican economy in the coming years. The ensuing prosperity may enable the political leadership to avoid making difficult agricultural decisions; yet on the other hand it might create a relaxed atmosphere such as would enable the agricultural dilemma to be honestly faced and gradually resolved.

1.The Success Story

From the early seventies onward there has been a great deal of public discussion of farming and its problems in Mexico, partly because the nation has become aware that all is not well in this vital sector. It was disquieting for the nation to find in the 1970s that the country had drifted into the position of becoming an importer of basic foodstuffs; likewise the social conscience had been shocked by revelations of the persisting poverty of many rural people. It had long been assumed that Mexico had sufficient land to feed itself and to provide a substantial quantity of agricultural exports. Technical progress would assure an ever-continuing growth of output and of farmers' incomes. The discovery that all this was no longer inevitable produced a rather violent reaction in public opinion—and a frantic search for remedies.

By the nature of the political process most of these remedies have been short-term in character, but agriculture is by its nature an activity in which situations change slowly and investments take many years to mature. To bring about durable improvements it is necessary to take a long view, to diagnose the more basic causes of present shortcomings, and to devise policies to take effect over a span of years. The present study is oriented toward that objective. Its point of departure is the current malaise in the farm sector, its time horizon is that of 1990, and it attempts to address the question: what sort of agriculture will be appropriate when there are nearly 100 million Mexicans, all more prosperous and demanding than the 68 million of 1980. Above all, this study seeks to suggest what measures need to be initiated in order to satisfy the social and economic aspirations of the farm people of the 1980s and beyond.

Every Mexican would be glad to have an efficient and progressive farm sector, producing the nation's requirements of food and agricultural raw materials and making a substantial contribution to merchandise exports; a farm sector utilizing to the fullest possible extent the land resources of the Republic; a farm sector exploiting for irrigation purposes a high proportion of the available water supplies; a farm sector pressing the frontiers of research in order to develop more productive plants and animals and to discover more

[1]

skillful ways of managing the various soil types in the differing climatic zones; and a farm sector disseminating among farmers of all categories the technical know-how to accelerate the adoption of modern management practices. This is not just a perfectionist's pipe-dream. It is a perfectly legitimate aspiration. Indeed in the light of Mexico's present demographic explosion the realization of something approximating this picture has become a national necessity.

Every Mexican would also wish to have his farm people earning fair rewards for the work they perform and would hope for the progressive elimination of unemployment and underemployment in the countryside. He would want farmers and their families to enjoy standards of living comparable to those of urban workers at equivalent levels of skill, thereby bringing to an end the exploitation of the farm sector by the city sector and removing a major injustice which constitutes a cancer in the social structure of the nation. This too is a legitimate aspiration, even if it has been achieved in very few countries, because it is a goal which could be attained by intelligent economic and social engineering, thereby making available to farm people the equipment and tools they need for their march toward prosperity.

Neither of these aspirations can be fulfilled overnight nor indeed within the span of a single presidential term. Land and animals have rhythms of their own which, by the inherent nature of the agricultural production process, have a much longer time span than the processes of manufacturing industry, so that problems which superficially appear to be short-term in character require in reality long-term measures for dealing with them. This does not mean, however, that agriculture is condemned to be the slowest performer among the sectors nor that farm people cannot begin to catch up with their urban brethren in prosperity. It means that the efforts to attain the goals may have to be more far-reaching and certainly more sustained than has hitherto been generally recognized. Rather than concentrating on the output of next year to the exclusion of all else, it will be prudent to raise our eyes toward more distant horizons and prepare for fulfilling those major aspirations before 1990.

A MAGNIFICENT PERFORMANCE

There may be some danger that the widespread attention focused on the present shortcomings of Mexican agriculture could cause people to forget the remarkable achievements of the farm sector, achievements which have few parallels anywhere in the world. To be understood adequately the problems of agriculture need to be viewed in the perspective of agriculture's own history, and in the case of Mexico that history over recent decades gives cause not for shame but for pride. Mexican farming has shown a truly astonishing capacity to meet the demands made upon it by an exploding population and at the same time has, through its exports, earned an important proportion of the foreign exchange needed to pay for the import of the machinery and raw materials for industrial expansion.

During the first decade of the century, crop output was increasing at about four percent per year.* This was followed by a sharp fall during the revolutionary years, leading to a period of slow recovery (1920–40), although even by 1940 production had not quite regained its 1910 level. From 1940 rather suddenly crop production started to climb steeply and maintained its dizzy rise until the middle sixties. During this remarkable 25-year period, production more than quadrupled, rising at 5.7 percent per year (Table 1.1). Since 1965, as we shall see, the rate of growth has been much more moderate.

TABLE 1.1

GROWTH IN QUANTUM OF AGRICULTURAL PRODUCTION:
1900 TO 1965

Period	Crop Products	Livestock Products	Total
	Percentage Increase per Period		
1900–1910	47.0	...	
1910–1920	− 26.9	...	
1920–1930	8.8	...	
1930–1940	33.1	46.0	40.5
1940–1950	75.3	55.9	63.9
1950–1960	58.4	25.5	41.3
1960–1965	44.3	27.7	37.1
1940–1965	300.7	147.1	217.5
	Compound Annual Rate of Growth		
1900–1910	4.0	...	
1910–1920	− 3.0	...	
1920–1930	0.9	...	
1930–1940	2.9	3.8	3.4
1940–1950	5.8	4.5	5.1
1950–1960	4.7	2.3	3.5
1960–1965	7.7	5.0	6.5
1940–1965	5.7	3.7	4.7

SOURCES: CROP PRODUCTS
1900 to 1930: Index calculated by Angulo, H. G., 1946.
1930 to 1950: Statistical Office, decennial farm censuses, volume data valued at 1950 price of each product, as published by the Department of Agriculture.
1950 to 1965: Department of Agriculture, Bureau of Agricultural Economics, annual crop reports, volume data valued at 1950 and at 1970 prices; arithmetical average of the two results.
The Department of Agriculture's 1950 output value was adjusted upward by 11.5 percent for believed underestimation of corn and beans; see Banco de México, *Proyecciones de la oferta y la demanda de productos agropecuarios en México a 1965, 1970 y 1975*, México, 1966.

LIVESTOCK PRODUCTS
Author's estimates as explained in Appendix. Prices as for crop products.

---

*Numerous problems arise in the interpretation of Mexico's agricultural statistics. In order not to load the text with technical discussions, the topic is treated separately in the Appendix, where reasons are given for preferring one set of figures to another.

For livestock products, the calculation of output has been made only for farm census years (plus an estimate for 1965), since, owing to the dubious nature of the data, it would be pretentious to estimate annual output changes on the basis of the unreliably recorded annual changes in livestock numbers.† It will be noted that livestock production expanded at only a moderate pace between 1930 and 1960—in fact it suffered a severe setback from the foot and mouth disease epidemic in the late forties. Rapid expansion was delayed until the sixties.

Agricultural production as a whole, combining crops and livestock, rose extremely rapidly from 1940 to 1950, more moderately during the fifties because of the livestock recovery problem and again very rapidly in the first half of the sixties. Much praise has been accorded to China for having doubled her food production in 27 years[1], but Mexico's food production doubled in only 15 years (1950–65).

The entire period was characterized by intense agricultural activity in many different parts of the Republic—opening up of new land, putting ambitious irrigation projects into operation, cultivating new crops on a large scale, and producing high-yielding varieties of crops—a startling success by plant breeders. Seen in historical perspective, those 25 years may legitimately be referred to as the golden age of Mexican agriculture. What factors made possible this remarkable performance will be examined presently, as will the events which occurred after 1965, but it may well be that for a variety of reasons such an expansion was unique and is unlikely to be experienced again.

Mention was made earlier of Mexico's high international standing when rates of agricultural growth are compared among countries. An illuminating comparison has been presented by Dovring utilizing FAO agricultural production index numbers (see Table 1.2). Although the base period of the comparison is slightly earlier in time, the index numbers broadly cover the era under discussion. Table 1.2 shows the relative stagnation of farming in the formerly renowned agricultural country, Argentina, and the outstanding superiority of Mexico's performance. (The discrepancy between the figures for Mexico's farm output shown in Tables 1.1 and 1.2 is due largely to the underestimation of livestock production in the latter, an error of long standing.)

By a curious coincidence, or perhaps it was not entirely coincidence, Mexico's agricultural explosion occurred simultaneously with the beginning of her demographic explosion. Figure 1.1 shows the two trends from the year 1900 to 1980: production growing faster than population from 1900 to 1910 and again much faster than population from 1940 to 1965.

Probably, however, the acceleration in the population growth rate from

---

†See the Appendix for an explanation of the methodology used in estimating the output of livestock products.

[1]Aziz, S., ed., *Hunger, Politics and Markets,* New York University Press, New York, 1975, p. 101.

TABLE 1.2

**INDICES OF GROWTH OF AGRICULTURAL PRODUCTION IN SELECTED
LATIN AMERICAN COUNTRIES**
(average for 1962–65)

Country	Index (1934–38 = 100)
Mexico	314
Colombia	236
Brazil	210
Peru	202
Chile	164
Uruguay	135
Argentina	128

SOURCE: Dovring, Folke, 1973, p. 38.

1930 onwards did provide an important stimulus to the farm sector, but, inasmuch as a large element in the new demographic situation was a reduction in the infant mortality rate, this did not show up in terms of an increased food demand until the additional infants had put on a few years; consequently the impact on agriculture was delayed until the early forties.

The major expansion took place in the basic foods required by the Mexican people. The crop area devoted to corn and beans expanded in proportion to the overall expansion and continued to account for over 80 percent of the area under crops. Wheat was beginning to become a basic crop because the process of urbanization began to convert a part of the urban population into consumers of wheat rather than of corn, although not until the fifties did wheat production expand sufficiently to reduce and finally eliminate the need for wheat imports.

Because the forties happened to be the war years, a strong demand for Mexico's exports led to a rapid expansion of cotton, tomato, and sugar production and, to a lesser extent, of coffee and other tropical products. Although the export stimulus was marginal in terms of acres devoted to export crops, less than seven percent of the total harvested area, its impact was much more significant and widespread as is often the case with marginal changes.

The proportion of the total agricultural output (measured in terms of value) devoted to exports increased marginally between 1940 and 1950 from 8.5 percent to 9 percent, cotton replacing the cattle lost through the foot and mouth epidemic, but in the following decade it jumped to 14.7 percent[2]. The years of the agricultural miracle were the period when the Mexican economy, for the first time since the Spanish invasion four hundred years earlier, ceased to be principally dependent on minerals. While in the year

---

[2]Centro de Investigaciones Agrarias, *Estructura agraria y desarrollo agrícola en México,* México, 1970, vol. 1, pp. 248–49. Rather higher percentages are given for 1950 and subsequent years in M. Rodriguez Cisneros, *Características de la agricultura mexicana,* Banco de México, p. 148.

FIGURE 1.1

INCREASE IN POPULATION AND FARM OUTPUT: 1900 TO 1965

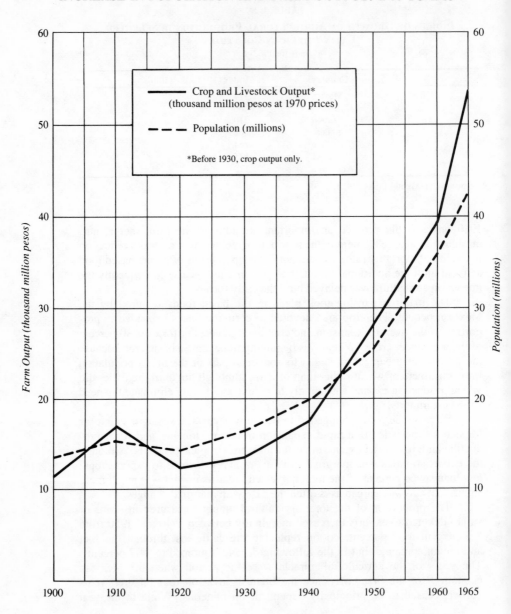

SOURCES: *Population*—decennial censuses with official revisions of the data for 1960 and
1965.
Agricultural Production—crop products 1900 to 1930 from Nacional Financiera,
S.A. *La economía mexicana en cifras,* linked in 1930 to Secretaria de Agricultura y
Recursos Hidráulicos, Dirección General de Economía Agrícola, *Consumos
aparentes de productos agropecuarios, 1925–1976;* livestock products are author's
estimates.

1939 minerals still accounted for 65 percent of total merchandise exports and agricultural produce for only 28 percent, by 1950 the situation had been radically transformed, agriculture's share having risen to 55 percent and minerals' falling to 33 percent.[3]

During the fifties and sixties agriculture continued to account for about half of Mexico's exports, but slowly another basic transformation was preparing to take place, the transformation into an industrial economy. In the year 1974 Mexico crossed an important threshold when her exports of manufactures finally came to account for more than half (54.3 percent) her total exports; farm products had fallen to 38.6 percent, and minerals to 7.1 percent.[4] With this last and final change, Mexico terminated her foreign trade dependence on primary products (other than oil) and joined the ranks of the newly industrialized nations.

The role of agriculture has also declined as a contributor to the country's Gross Domestic Product. A number of conflicting figures are available in various publications[5], for instance the contribution of crops and livestock products in 1940 to Gross Domestic Product ranging from 19.1 percent to 23.9 percent, and in 1965 from 13.6 to 17.2 percent, but all agree that the percentage fell during this period. In other words, in spite of the dramatic expansion of agriculture the other sectors were expanding even more rapidly. This, of course, replicates the experience of all countries whose economies are becoming more diversified, but among the few countries whose Gross Domestic Product was growing as fast as that of Mexico none recorded such a well-maintained share by agriculture. In fact the dynamism of farming fell little short of the dynamism of industry and commerce.

During the golden years of its expansion, Mexican agriculture was also making a real contribution to employment even though in the industry and services sectors the volume of employment was increasing more rapidly. The bringing into cultivation of new lands, the rapid extension of irrigation, the introduction of more labor-intensive crops, and the increase in livestock numbers, all added up to a requirement for more manpower which was not, at least up till 1960, offset by the trend toward mechanization. Significant in this respect was the contrast between what happened in Mexico and what occurred in almost all the other developing countries. In these latter countries during the forties and fifties the population recorded as "economically active in agriculture" was also increasing, but agricultural output and consequently its manpower requirement was expanding very slowly, so that what was augmenting most was the number of unemployed and underemployed farm people; whereas in Mexico the dynamic need for farm manpower enabled

[3]*Problemas Agrícolas y Industriales de México*, vol. 4, no. 3, México, 1952, p. 201.
[4]Secretaria de Programación y Presupuesto, Dirección General de Estadística, *Agenda Estadística*, 1975, pp. 158–59.
[5]See *Problemas Agrícolas y Industriales de México*, vol. 4, no. 3, 1952; Rodriguez Cisneros, M., 1974; Navarrete, Ifigenia M. de, ed., *Bienestar campesino y desarrollo económico*, Fondo de Cultura Económica, México, 1971; Instituto de Investigaciones Sociales, *El perfil de México en 1980*, Siglo Veintiuno Editores, México, 1970; and Banco de México, S.A., *Informe Anual 1974*.

most of the increase in the farm labor force to obtain employment, at least during parts of the year, with the result that un- and underemployment almost certainly did not increase and may well have declined by 1965. (This impression cannot be quantified because of the immense difficulties in interpreting the agricultural population statistics and in estimating the effective volume of farm employment.)

The agricultural explosion likewise generated a remarkable improvement in incomes, and although, as in the case of employment, this is difficult to measure with any precision, certain indicators may be cited as significant. First of all, the gross agricultural output per person economically active in farming increased in volume terms (i.e., at constant 1950 prices) by about 50 percent between 1940 and 1960. Moreover, net agricultural output, after deduction of seed and feedstuffs grown on farms, increased slightly more than the gross output. To arrive at the amount available as income and profit, we have to make a further deduction for the cost of inputs purchased from outside the farm sector, a sum which probably increased only moderately and not faster than net output between 1940 and 1960, since the great increase in the use of machinery and fertilizers occurred later. It is therefore probably fair to retain a figure of roughly 50 percent as representing the improvement in per capita farm income during the 1940–60 period. Admittedly this constitutes a gain of only 2 percent per annum, but that was far superior to the contemporary performance of most other developing countries.

This was also a period in which a considerable amount of redistribution of agricultural incomes was taking place. Progressively, the large estates were being reduced in number and size, partly through the creation of ejidos and partly through landowners' own action in selling off parts of their properties to smaller-scale farmers. Between 1940 and 1970 what are euphemistically described as "small farm properties," but which 45 years ago were anything but small, reduced their total area by 58 million acres (or by nearly a quarter) while ejidos expanded by 78 million acres (the difference being due to the increase in the area covered by the censuses.)* Between 1940 and 1970 the area farmed in properties of over 250 acres declined from 230 to 156 million acres and the great majority of those that were left were ranches of low-grade pasture. The whole subject of the changes that have occurred in farm sizes will be discussed in some detail in Chapter 8, but these few figures indicate the magnitude of the land redistribution process. And, since the quantity of land available to a farm family is a major determinant of farm income, this massive redistribution of land clearly contained an income distribution component.

Nor was this all. In the forties and fifties, due to the opening up of new areas to cultivation, a very significant internal migration was taking place

---

*The Agrarian Reform Act of 1915 created two new classes of land tenure; (1) communities *(comunidades)*, a resuscitation of communal ownership on lands where this existed prior to their expropriation to private farmers in the nineteenth century; and (2) *ejidos*, a similar form of tenure applied to lands taken from landowners after 1915. By the seventies the ejidos had come to possess nearly half the farm land of Mexico. Their members, *ejidatarios*, hold their plots, *parcelas*, in usufruct. For a full discussion, see Chapter 7.

within the Republic, especially from states of the center and the south to those of the north and northwest. With few exceptions the people who participated in these migrations bettered themselves and attained levels of income far superior to what they had earned in the villages whence they came. From a social point of view, it is perhaps regrettable that the settlement movement did not take place on an even larger scale because this would have relieved more considerably the demographic pressure in the zones of agricultural poverty and would have enabled a larger number to participate in the progress being achieved in the new areas.

The areas in which little or no income improvement occurred were those where ecological circumstances were so harsh as to impede the adoption of modern farming techniques.

It has been said that the opening up of new lands and the building of irrigation works intensified the income differences in the Mexican countryside just as some writers have accused the green revolution of accentuating the contrast between haves and have-nots in India and Pakistan. A more objective description of what was happening between 1940 and 1965 in Mexico would identify three distinct processes occurring simultaneously: first, a descent of formerly large landowners from the top to a middle income group; secondly, an ascent of many poor people, especially the migrants, from the bottom to various middle income groups; and thirdly, a continuation in poverty by those who remained living in the hostile and marginal agricultural areas. This, of course, constitutes a very broad generalization—in real life there existed an infinity of graduations and subcategories—but it portrays a very normal societal process of the kind which occurs when a nation has marched the first few miles along the road toward generalized human welfare. It is also important to remember that Mexico was one of the few developing countries in which the agricultural sector was playing a prominent role in this march.

Furthermore, the farm sector was making a powerful contribution, albeit indirectly, to the creation of employment and incomes in other sectors of the nation's economy. It is a phenomenon which can be observed in all parts of the world that a flourishing and efficient agriculture is a precondition for advances in industry and general welfare, because a country which is industrializing itself needs to import a great quantity of raw materials, machinery, and equipment which it does not possess, and it cannot afford at the same time to import food. In such circumstances the task of agriculture is to provide the food required by the expanding urban population and to contribute exports which pay for machinery imports until such time as the role of paying for imports can be taken over by manufactures and other nonagricultural commodities. If agriculture fails to fulfill this mission, the import of equipment for industry has to be slowed down, industrial employment and output cannot rise rapidly, and the population as a whole cannot enjoy access to basic consumer goods and much needed social services. Many are the countries which have been unable to move forward rapidly for lack of dynamism in their farming sector. Mexico has been extremely fortunate in having had for a considerable period an agricultural sector sufficiently virile and dynamic to underpin her advances in the social and economic well-being of the population as a whole.

THE TECHNICAL COMPONENTS

An analysis of how and why the agricultural explosion occurred would show which among the factors mainly responsible can be expected to be available for the new agricultural effort needed in the coming years. A detailed investigation of what occurred would belong to a volume on the agricultural history of the period, but for the purposes of the present survey it should be sufficient to indicate the highlights of the transformation leaving to later chapters their application to the present situation and to future perspectives.

The most basic and important factor during the 1940–65 period was the expansion of the area in agricultural use, both for crops and for livestock. From the date (1930) of Mexico's first agricultural census until 1940 the rate of increase was moderate, but then suddenly the expansion began to accelerate, reaching a peak in 1966. Regarding the magnitude of this expansion, the statistical sources conflict, with the farm census giving an annual growth rate in harvested area of 2.4 percent (1940–60) and the Department of Agriculture one of 3.4 percent (1940–65), a discrepancy which is discussed in the Appendix. This large increase came partly from an extension of the arable area, partly from a reduction in the acreage fallowed, and partly from reducing the area of crops lost (through droughts, pests, etc.) prior to harvest, both these latter providing evidence of widespread advances in farming efficiency. Part of the area expansion was located in rain-fed areas, principally in the Gulf states, but part also in hitherto desert areas of the north and northwest which were being opened up to cultivation by irrigation works.

This indeed was a second major factor making possible the agricultural miracle: the area under irrigation doubled between 1940 and 1960, attaining a total of 8.6 million acres as a result of massive investments by the National Irrigation Commission (which were continued after this body was transformed in 1946 into the Ministry of Hydraulic Resources). The extent of the benefits conferred by irrigation of course varied from region to region, in accordance with local ecological conditions, but it may be estimated that on average the provision of water more than doubled the per acre yield of crops besides enabling certain crops, such as cotton, to be grown profitably—crops which had hitherto been confined to small areas of the best rain-fed lands.

A further component of technical progress was the wider use of material inputs, in particular improved seeds, fertilizers, pesticides, and farm machinery. Of these probably the most important during the 1940–65 period was the provision of improved seeds. The Rockefeller Foundation established in Mexico a plant-breeding center (later known as CIMMYT) which attracted several outstanding specialists and which by the late fifties had succeeded in producing several conspicuously successful varieties of wheat adapted to Mexican conditions (varieties which later provided the basis for the green revolution in India and Pakistan and other countries). They also achieved notable improvements in corn, other cereals, and some oilseed crops. The interest which these successes awakened among the more progressive farmers provoked them into seeking also to cover their seed requirements of other crops by imports from abroad.

The availability of water on a secure basis made it for the first time profitable to utilize chemical fertilizers on a substantial scale. Consumption of fertilizers expanded from 8,000 tons in 1950 to 346,000 tons in 1965. Moreover, the creation of large compact areas of irrigation, the so-called irrigation districts, made possible the use of farm machinery on a much larger scale than previously. The greater speed at which machinery could operate, compared with the speed of animal-drawn implements, enabled sowing and harvesting to be accomplished at much more optimal rates, and facilitated more precise control of sowing depths and spacing of seeds. Mechanization and the use of aircraft also made possible the widespread utilization of pesticides which contributed significantly to the improvement in crop yields.

The overall expansion in crop production was further aided by the progressive introduction of high value crops, especially in the irrigated areas, crops such as cotton, sugarcane, wheat, tomatoes, alfalfa, and later oilseeds and grain sorghums.

There were thus three main factors operating together to bring about the 1940–65 expansion of crop production: increase in harvest area (3.2 percent per year), increase in per-acre yields (2 percent), and changes in relative importance of the crops cultivated (less than 1 percent).

Because the livestock data are so deficient, it is impossible to present an analysis in any respect comparable to that for crops. As is well known, the essential components determining changes in the output of livestock products are first the number of animals in each category and secondly the changes in the productivity of those animals. Regarding the changes in livestock numbers, the figures in Table 1.3 would appear to be a reasonable choice from the confusing, unreliable, and contradictory data available. Since there are no census data for 1965, the final year of the agricultural boom, an overall figure for the year 1940 to 1970 has been added to the table. It may be noted that for hogs and poultry the major advances were recorded in the sixties, but for sheep these were in the thirties, and cattle and goats in the forties. The poultry industry was by far the most dynamic section except for an apparent pause in the fifties. Over the thirty-year period, the annual growth rate of poultry numbers was substantially higher than that of cattle and hogs.

It is impossible to cite any reliable data for changes in the output per head of livestock, i.e., their productivity, which would be a concept corresponding to the changes in crop yields per acre. The problems are explained in

TABLE 1.3

GROWTH IN LIVESTOCK NUMBERS: 1930 to 1970

(annual rates of growth)

Period	Cattle	Hogs	Sheep	Goats	Poultry
1930 – 1940	1.4	3.3	2.0	0.4	5.2
1940 – 1950	3.1	3.1	1.4	2.2	4.6
1950 – 1960	1.2	– 1.2	0.1	1.4	0.1
1960 – 1970	2.6	4.7	– 0.4	– 0.4	6.9
1940 – 1970	2.3	2.1	0.3	1.0	3.8

SOURCES: Statistical Office, decennial farm censuses.

the Appendix, and our tentative conclusions are reflected in the output figures in Tables 1.1 and 1.5.

ECONOMIC AND SOCIAL COMPONENTS

Up to this point only the physical and technical components of the agricultural miracle have been reviewed, but account also must be taken of economic and social changes affecting the farm sector. The fact that these are less easy to quantify and that many can be appreciated only in qualitative terms does not make them less significant.

Consider, for example, the economic stimulus afforded to farming by changes in prices of individual agricultural commodities, a matter which at first sight might appear capable of mathematical calculation. We can certainly tabulate the changes in "average rural prices" of individual crops and can calculate in more ways than one a weighted average index of agricultural prices as a whole. For instance, while the agricultural price index more than trebled during the inflationary forties, it merely doubled between 1950 and 1965, a rather smaller increase than that of wholesale prices in general during the same period. In real terms, therefore, agricultural prices declined somewhat.

This bald statement, however, masks very different price behaviors by particular crops. Thus, while the prices of a few commodities such as oranges nearly quadrupled, the price of others such as cotton hardly increased at all, and yet others such as wheat by a mere 50 percent. It also hides the important consideration that the fact that commodity A enjoys a stronger price rise than commodity B over a period of years does not mean that the production of A will necessarily be stimulated more than the production of B. It is necessary to see what has been happening during the period to the average and marginal costs of production of A and of B. The extent of the stimulus depends not on prices nor on costs but on the relationship between the two. Unfortunately, in the case of Mexico no representative data exist concerning production costs even of major crops, and therefore no measurement of the economic stimulus is practicable. Indeed, we are forced to approach the topic the other way round by saying that the fact that the cultivation of a particular cash crop was expanding rapidly in a certain region is a clear indication that its profitability was increasing relative to other crops capable of being cultivated in that region. A profit stimulus may occur even when a commodity price is falling in real terms provided that unit costs of production are falling faster.

A case in point is cotton, whose price was falling steadily in real terms throughout the 1940–65 period while the area under cotton increased persistently—the reason being that as a result of improved varieties and better cultivation practices, including the use of fertilizers and pesticides, production costs per bale of cotton were falling even more rapidly, enabling the incomes of cotton growers to increase. The one safe conclusion which may be drawn following this line of argument is that in the case of those cash crops which enjoyed a major area expansion, their profitability was undoubtedly increasing. But the same reasoning cannot be applied to subsistence crops whose value of production is governed not so much by market forces as by demographic pressures and the compelling need of the rapidly increasing farm population to feed itself.

The phenomenon of falling costs and increasing incomes characterized the farming in virtually all the "new" areas, new in the sense of newly opened rain-fed lands or newly established irrigation works, and constituted a major motivation for the agricultural expansion. One of the important questions which will be examined in later chapters is whether or not there are good prospects of this phenomenon occurring again on any significant scale in the years ahead.

Another area in which changes took place was in the concept and practice of farm management. For centuries in Mexico, as in other countries, agriculture had been almost exclusively subsistence farming and in consequence was universally regarded as a way of life rather than an economic activity. Gradually, as urbanization and industrialization grew, a few farmers began to concern themselves with supplying the needs of the cities. Yet, even as late as the 1930s this constituted a small fraction of the farm population; the remainder stood outside or barely on the fringe of the market economy, with the result that for them the concept of farm management had no relevant meaning. This was so not merely for the millions of mini-farmers but also for the majority of large landowners who regarded their estates more as prestige symbols than as economic instruments and who continued farming along traditional lines.

The advent of irrigation and modern cultivation techniques changed all this, and there began to emerge a new type of farmer, whether small proprietor or colonist or ejidatario, who regarded farming as a business operation requiring the application of management techniques. This became most strikingly evident in the northwest region where the entire pattern of farming was radically changed and the break with the past was most pronounced. But it also spread progressively to irrigation districts elsewhere, to plantations and to some livestock enterprises. A feature of this new attitude was the constant search for improvements, technical and economic: better seeds, more effective fertilizer mixtures, less expensive inputs, better methods of marketing. Inertia and conservatism were replaced by experimentation and dynamism. Those who could adopt these new approaches quickly outdistanced those who could not. In the northwest, the changeover was most widespread largely because such a great proportion of the farm population were immigrants from other parts of the Republic, who by the very act of removal had broken with the customs and traditions of the villages from whence they came.

All this meant that Mexican agriculture, or rather a section of it, acquired a new motivation for progress which expressed itself in the behavior of business-minded individuals who were receptive to new technologies and whose example slowly percolated downward. It was the spread of those attitudes, quite as much as the fact that irrigation and new techniques were becoming available, which brought about the golden age of Mexican farming. As always in the affairs of society, the human elements can be quite as essential and decisive as the physical and technical elements.

REGIONAL IMPACT

All these changes in the evolution of crop and livestock production have so far been treated at national level and the quantification of progress achieved has been in terms of the Republic as a whole. But Mexico is far from homogeneous: in reality there are many Mexicos, each with its distinctive geophysical environment and its differing human characteristics. While it would be instructive to make a more detailed examination of what was taking place region by region, even state by state, such a series of computations would probably do little more than illustrate what is already obvious, namely that the states in which it was feasible to introduce modern technology achieved quite extraordinary advances, whereas those, notably in the Altiplano, which had to continue cultivating marginal lands under conditions of insufficient and erratic rainfall made much less conspicuous progress. However, it may serve by way of illustration of this general proposition to present data from two contrasted states: Sonora, representing the irrigated northwest, and Hidalgo, representing the disadvantaged center.

TABLE 1.4

Agricultural Changes in Two Mexican States:
1950 to 1970
(1950 = 100)

	Sonora	Hidalgo
Crop Output	484	180
Harvested Area	211	135
Crop Output per Harvested Acre	230	133
Cattle Numbers	189	130
Poultry Numbers	649	138
Output of Livestock Products	240	161
Population	174	112
Farm Ouput per Capita	290	171

Source: Statistical Office, decennial farm censuses and censuses of population.

Between 1950 and 1970 the output of crop products measured at constant 1970 prices increased nearly fivefold in Sonora but by only 80 percent (much less than twofold) in Hidalgo. This was caused by a variety of factors, but mainly by an enormous increase in irrigation in Sonora, making possible a big gain in harvested area and in per-acre yields.

In Hidalgo already in 1950 most of the usable land was in cultivation so that the harvested area expanded by only 35 percent over the ensuing twenty years, very little of the total being irrigated. Cattle and poultry numbers increased very slowly. In these depressing circumstances it is not surprising to note a strong outflow of population from the state of Hidalgo, the total rising by only 12 percent (national average 87 percent) while Sonora's increased by 74 percent. The adjustment through internal migration enabled Hidalgo to register a modest increase in farm output per head, whereas in Sonora, despite immigration, the per capita output increased nearly threefold.

Other states of the Republic experienced different impulses to development: for instance, in Zacatecas a big increase in cropped area but a very small increase in per acre yields. In some cases distinctive trends occurred in different regions within the same state, for instance in Oaxaca a large increase in cultivated area and yields in the Papaloapan basin but virtual stagnation in the Alta Mixteca. These examples should emphasize sufficiently that although for many purposes we are obliged to discuss changes in terms of national averages we should never forget the tremendous contrasts from region to region in the character and tempo of development.

A CHANGE OF RHYTHM

Around 1965–66 the golden age of Mexican agriculture came to an end. For twenty-five years the country had enjoyed unprecedented farming progress at an incomparable speed, but like so many good and splendid things it was implacably fated to change from a glorious gallop into a temperate trot. What happened can be seen in Figure 1.2, showing year to year movements in the value of crop production (at constant 1970 prices). Instead of continuing upward at the 1940–65 rate of growth (dotted line in graph), the expansion slowed down markedly, experienced setbacks in 1967 and 1969, and altogether achieved over the years 1965–80 an average annual growth rate of not quite 2.6 percent compared with 5.7 percent in the golden years of 1940–65. Among the important crops the output of corn, beans, sugarcane, and bananas hardly increased at all, while cotton, copra,and henequen declined. Only wheat, sorghum, and oilseeds showed good increases. The growth in output of livestock products also apparently slowed down.

A growth rate of 2.6 percent can hardly be considered satisfactory when Mexico's population was increasing more than three percent and her internal demand for farm products at close to five percent. Inevitably by the end of the period agricultural products were being imported on a substantial scale for the first time in many decades and agricultural exports were shrinking.

It would have been important, therefore, to ascertain what had gone wrong and where and why, and, having identified the problems, to adopt policies for dealing with them. But this was not the initial reaction in government circles. Senior officials in their public statements put the blame fairly and squarely on the weather, and this explanation continued to be given major prominence up to and including the year 1980, even though there had not been a generalized weather disaster since 1969. Of course it is quite easy to refer to a drought in Durango and Zacatecas, to floods in Tamaulipas and Veracruz, or in some luckless zones to both droughts and floods occurring in the same year; but there has not been a single year in Mexico's history when some part or other of the Republic did not experience a drought or a flood. These are the normal hazards of farming in tropical and subtropical climates, and there is no evidence that these afflictions were above average in impact during the period under review. However, it is much easier to blame the weather than to report to ministers unpleasant facts which they may not want to hear, and this very normal human weakness is likely to display itself more conspicuously where

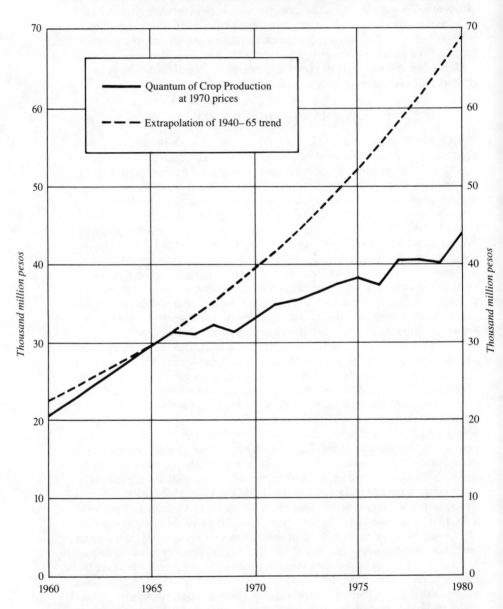

FIGURE 1.2

QUANTUM OF CROP PRODUCTION: 1960 TO 1980
ACTUAL (AT 1970 PRICES), AND 1940–65 TREND (5.7% PER YEAR)

Thousand million pesos

——— Quantum of Crop Production
at 1970 prices

– – – Extrapolation of 1940–65 trend

SOURCE: Secretaria de Agricultura y Recursos Hidráulicos, Dirección General de Economía Agrícola, *Consumos aparentes de productos agropecuarios, 1925–1976,* and annual reports thereafter for quantities. Average rural price in 1970 is used for each crop.

governments have engaged their reputations in successfully guiding the development of the national economy. The greater the extent of State intervention the greater the temptation to misrepresentation on the part of bureaucrats at all levels in the hierarchy, the temptation being to please the boss by pretending that any setback is unimportant, unavoidable, or both.

Moreover, in order to be able to detect the factors responsible for a change of trend in a production process as complex as agriculture it is necessary to have extensive, reliable, and up-to-the-minute statistical reports covering a wide range of relevant topics. Few countries have such a sophisticated service, and Mexico is not one of them. What usually happens is that as the months and years go by first one and then another fragment of information becomes available until progressively it becomes possible to piece together a picture of what is happening in the countryside. It would have required a remarkably perspicacious person to diagnose correctly agriculture's difficulties as early as 1971 or 1972, whereas by 1976 the situation had become extremely clear.

To put the problem in a single sentence we may say that every single one of the components of the successes of the golden era was changing after 1965 into a component of adversity, some more violently, others less so; but in combination their influence was becoming negative and inevitably applied a brake to the rate of expansion of farm output. The agricultural sector was beginning to face a series of physical, technical, economic, and social limitations which would make a continuation of growth much more difficult than previously, and which would require in consequence a rethinking of agriculture's place in the nation's economy and of the policies suitable for guiding the farm people toward their new horizons. Since such a rethinking is the subject matter of the present study, most of the later chapters are devoted to diagnosing the nature of these various limitations and to exploring alternative ways of overcoming them. Therefore, all that seems necessary at this stage is to list very briefly what the "components of adversity" are so that the reader may become aware of the magnitude and variety of the problems which have to be investigated before we can feel equipped to formulate policy recommendations.

It will be recalled that in the 1940–65 expansion of crop production the major physical components were twofold: an expansion of the harvested area and an increase in per acre yields. As regards the harvested area this reached a peak of 38.8 million acres in 1966 which was not regained till 1977. This change of trend to zero growth in area was in itself sufficient to account for most of the decline in the rate of growth of crop output. Furthermore, the additional irrigation works completed in the later sixties and the seventies were very modest compared with the works of previous decades. Fortunately, the other significant component, namely yield, continued its upward trend. Over a period as short as a decade it is impossible, because of the year-to-year fluctuations, to measure accurately the yield trend, but an examination of the performance of the principal crops suggests that their per acre yields have been increasing at a rate not much below 2 percent per year, which is only slightly inferior to the average of the previous decades. Finally, some contribution to the growth in the real value of crop output appears to have been derived from an improvement in the crop-mix, more specifically the output of high value fruits and vegetables increased proportionately more than the output of some low value annual crops. All in all, the data now available indicate that Mexican agriculture has come to a point of real difficulty in finding regions in which the crop area can be expanded but still is contriving by means of better farming practices and by shifting to more remunerative crops to improve the volume and value of the per-acre output.

In the livestock sector the information is hardly sufficient to make reliable judgments possible. From 1965 to 1980 the annual growth rates appear to have been close to the long-term average (but see Appendix). As for the trends in the productivity of these various classes of farm animals, there is no reason to suppose any decline: output of fodder crops has increased more than that of food crops while modern management methods have continued to spread. Inasmuch as both numbers and productivity have continued at more or less their previous rate of upward trend, it may be tentatively concluded that the output of livestock products has been growing at around 3.6 percent per annum, possibly a little more, but in any case has done better than crop products.

If the physical and technical impact of agricultural growth ran into difficulties during these ten years what had been happening to the economic and social components? Had their positive influence been sustained or had it likewise declined? Certainly, the stimulus from the side of prices persistently weakened in the sense that agricultural prices measured in real terms (i.e., the relative movement between the prices of things farmers sell and the prices of the things they buy) became more and more unfavorable, and not until late in 1973 and early in 1974 was this adverse trend corrected and then only partially. Moreover, the previous trend toward reductions in costs of production of many important cash crops, especially those cultivated in irrigated areas, was slowing down and in some instances came to a halt, partly because many of the major conquests in technification had already been achieved, partly because of the emergence of technical difficulties in the management of the irrigation districts, partly because of the slowing down in the completion of new irrigation works, and partly because of the inflationary rise toward the end of the period in the prices of various agricultural inputs such as fertilizers, pesticides, machinery, and petroleum products. Altogether, therefore, the economic stimuli to the maintenance of agricultural growth were becoming progressively weaker and weaker.

Similarly unfavorable changes were likewise occurring in the realm of farm management. Earlier it was seen how the bringing of business efficiency into farming had been a direct result of the massive increase in irrigated cultivation during the forties and fifties, there being little scope for management improvements in the traditional zones of dryland corn and bean farming. The marked slowing down in new irrigation activities meant a corresponding slowing down in the extension of modern management to new areas. Another factor was the impact of land redistribution which quickened its pace after 1965. Whereas before 1965 most of the land transferred from large land-

TABLE 1.5

**TRENDS IN CROP AND LIVESTOCK OUTPUT, 1950 TO 1980;
AND PROJECTIONS, 1980 TO 1990**

Year	Crops (censuses)	Crops (Dept. Agr.)	Livestock (author)	Total (censuses)	Total (Dept. Agr.)
	Value of Output at Current Prices (in millions of pesos)				
1950	5,141	6,346 [a]	6,608	11,749	12,954
1960	14,396	16,399	14,165	28,561	30,564
1965		29,487			
1970	24,683 [b]	33,149	29,005	53,688	62,154
1975		75,275	61,100		136,375
1980 [c]	. . .	220,000	175,000	. . .	395,000
	Value of Output at 1970 prices (in millions of pesos)				
1950	10,745	13,219 [a]	14,909	25,654	28,128
1960	18,053	20,571	18,808	36,861	39,379
1965		29,926	23,391		53,317
1970	24,683 [b]	33,149	29,005	53,688	62,154
1975		38,144	33,954		72,098
1980 [c]	. . .	44,000	39,640	. . .	83,640
	Average Annual Growth Rates at 1970 Prices (in percentages)				
1950 – 1965		6.3	3.1		4.7
1965 – 1980		2.6	3.6		3.1
Projections:					
1980 – 1990		3.0 [d]	4.0 [d]		3.5 [d]

SOURCES: Statistical Office, decennial farm censuses; Department of Agriculture, Annual Crop Production Reports; author's estimates of volume of livestock output, and average farm prices (as published) of livestock products.

[a] Corn and bean output upvalued; see Table 1.1.

[b] Adjusted upward from 22,084 million, for adverse climate that year.

[c] Preliminary data.

[d] Projections by author.

owners to ejidos had been of good quality (previously idle but capable of becoming productive in the hands of its new cultivators), the supply of this type of land was becoming exhausted so that after 1965 increasingly bad quality land was transferred. The former type of transfer had increased land productivity whereas the latter was entirely neutral in its effects. In certain instances when land for transfer was in short supply, well-managed arable lands and parts of well-managed livestock farms were distributed to ejidos with a consequent fall in the quality of management and in output.

Yet another factor was that many of the "small proprietors" who, during the golden age, had invested heavily in increasing the productivity of their farms, now cut back drastically on their investment programs. On the one hand they were being subjected to an intense cost-price squeeze which strongly discouraged launching out into expenditure on new plant and equipment, new land clearing and leveling, new water-pumping for irrigation and livestock, new storage barns, new machinery. On the other hand they were being increasingly subjected to verbal attacks by politicians and to physical invasion by organized activists which created a climate of insecurity of tenure and largely brought to a standstill the undertaking of long-term agricultural improvements. The adverse effects of these negative investment decisions will become far more significant as time goes by. If the deterioration in agriculture's economic and social environment be added to the physical and technical difficulties noted above, it is not surprising that the golden age gave way to a bronze age in which problems accumulated and the growth rate of farm output fell seriously below the growth rate in demand.

PALLIATIVES AND SOLUTIONS

Long before this diagnostic report could be pieced together, the government was forced into action by the stark fact of food shortages and the need to permit massive importations of cereals, oilseeds, and other foodstuffs. Publicly it might be convenient to continue blaming the weather, but privately it began to be recognized that it would be imprudent to expect beneficent changes in climate to solve the food supply problem. Quite naturally, as is so often the case when governments are faced with an emergency situation, the authorities had recourse to a series of uncoordinated measures which had the merit of being able to be implemented quickly whether or not they were medicines appropriate to the illness. Furthermore, most of them were measures possessing a useful element of popular appeal without having any awkward political content.

The first important move was to decree increases in the guaranteed farm prices of a number of basic products. This step was welcomed by the farming community as providing immediate alleviation of their difficulties even though, as in all farming communities, it was complained that the new prices were still not high enough to reflect increases in costs, and that in many cases the price relationships between basic crops had become distorted.

A second move was to authorize and indeed order a tremendous increase in bank lending to the farm sector. The total volume of agricultural credit rose from 4,000 million pesos in 1970 to over 30,000 million in 1979, the great bulk of the additional funds being made available in the form of crop credit. This dramatic move, which was given widespread publicity, of course brought immediate benefit to farmers by augmenting their cash reserves for the purchase of seeds, fertilizers, and consumer goods, and particularly benefitted ejidatarios to whom virtually all the increased facilities were directed.

A third move was to launch a campaign for increasing the cultivation of basic crops, especially corn and beans but also to a lesser extent wheat and rice, and by this means quickly reduce the need for cereal imports. To some limited extent this could be achieved by cutting down the areas lying fallow, provided the rains arrived at dates which permitted an extension of sowings, and to some extent it was achieved temporarily by reducing the area of other crops such as cotton, and to some extent by expropriating some of the cattle ranchers' better quality pastures and turning them over to ejidatarios for crop growing.

The government also organized a rapid expansion of the agricultural extension services. Luckily it had previously embarked on a program of multiplying the number of higher-level agricultural schools and of agricultural colleges with the result that around 1974 a first crop of qualified field workers became available, with a still larger outflow thereafter. Although initiated during the emergency, this massive expansion of the extension service can be expected to act as an extremely positive factor in the longer term.

Yet another measure, also more long-term in its impact, which the government of President Echeverría adopted was a campaign for the collectivization of ejidos. With the country faced with a low level of technical know-how among the great majority of ejidatarios and the obvious difficulties of introducing modern farming practices on mini-sized plots, much might be gained in terms of output (so it was argued) by pooling the land resources of the ejido and operating them as a single unit under the guidance of a management committee and one or more trained agronomists. The increase in the number of extension officers would facilitate this reorganization, which was also supported by modifications in the credit policy of the government's Rural Credit Bank. Particular encouragement was given to collective livestock enterprises in ejidos and to crop collectives in new settlement areas.

These initiatives in respect to prices, credit expansion, basic crops, extension services, and collectives by no means complete the list of measures adopted by the government to promote a fair deal for the farm sector, but they have been singled out here as an indication of the main lines of policy reorientation adopted for dealing with what was widely regarded as a major internal crisis. In later chapters a more detailed analysis will be presented of these and various other supplementary initiatives in order to distinguish their short-term and their longer-term effects and to determine the extent to which they deal with the basic components of adversity.

Without any doubt, the crisis which built up within the farm sector since 1965 is not of a short-term transitory nature. It is deep-rooted in the technical apparatus and social structure of Mexican society. It is partly a physical problem and partly a human one, partly technical and economic, and partly institutional and social. It is an extremely complex problem, not solely because the farm sector has its own peculiar complexities, but also because, let it be repeated, there is not just one Mexico to which a uniform package of policies can be applied—there are many agricultural Mexicos, each with its own idiosyncrasies, physical and human; and programs which may succeed in one region may entirely fail in another. This diversity of circumstances will need to be reemphasized again and again in the course of the present study in order that the recommendations which may emerge can have practical applicability on the various types of farms and among the character-variegated farmers. One thing, however, remains certain: these farming problems are profoundly important not only for agriculture but for the whole nation, and a better understanding of them, leading to realistic revisions of policy, could decisively influence Mexico's economic progress in the years ahead.

2. Demand Prospects

As noted in the previous chapter, agricultural production during the "golden years" was increasing faster than the nation's demand for farm products, making possible a significant expansion of agricultural exports. However, since 1965 production has been growing more slowly than domestic demand with the inevitable result that agricultural exports declined and imports have increased. Since it is generally considered that this state of affairs is bad for Mexico and should if possible be rectified, any discussion of the future tasks of the farm sector needs to be prefaced by an examination of the calls likely to be made on it, an examination of whether the rate of growth of demand in recent years has been exceptional or whether it may be expected to persist. The present chapter, therefore, is devoted to an analysis of the various aspects of the demand for food and non-food agricultural products.

Demand for farm products has two elements: domestic demand, that is the quantities bought and consumed within the country, and foreign demand, the quantities purchased from Mexico by buyers in other countries. In considering domestic demand, some of the more important factors are: the rate of growth of population, the rate of growth of private consumption expenditure, changes in income elasticities of demand for farm products, changes in price elasticities of demand, changes in the relative prices of partly substitutable commodities, changes in food habits, and changes in the demand of various Mexican industries for agricultural raw materials (which in turn are affected by the domestic and foreign demand for products containing those raw materials). The factors influencing foreign demand include: the demand of importing countries for primary agricultural products, their demand for finished or semi-finished manufactures containing agricultural raw materials, the capacity of other countries to compete with Mexico in the export of these various commodities, and the extent to which the Mexican government either stimulates exports by means of direct or indirect subsidies or restrains them by means of export taxes and other disincentives. All these, as well as other more general influences, have to be taken into account when trying to evaluate future trends.

Demand projections, to be meaningful, should relate to a specific term of years. In such an exercise it is desirable to avoid extremes. A very short period would be inappropriate because agricultural changes can be brought about only slowly, while a long period would introduce too many uncertainties in the quantification of the demand components. Perhaps a middle course points to the selection of a ten-year period, say from 1980 to 1990, as reasonable for the present study; and the same decade will be utilized for other forward estimates in later chapters. This does not mean that we are trying to pinpoint any particular year either for the beginning or the end of our period; in these matters the major trends become modified gradually, and our conjectures are likely to be equally valid for, say, 1982 to 1992 or for three-year averages such as 1979–81 to 1989–91.

RECENT TRENDS IN DEMAND
FOR FARM PRODUCTS

In order to look forward intelligently it is necessary to begin by looking backward. A great deal can be learned from a study of what has been happening to demand in recent decades and from distinguishing the factors which influenced the observed trends. This does not, of course, imply a procedure of linear projections, because, while some of the operative factors may continue to be present, others will be changing. What it does is to offer an opportunity for establishing a more rational foundation for a projections exercise.

In Mexico, as in other countries, there are two possible sources of information about past levels of consumption: survey of family expenditure and consumption, and estimates of available national supplies. For reasons to be mentioned presently, the family survey data are quite inadequate for the present purpose, and we are left therefore with supply data at national level. These supplies are composed of national production plus imports minus exports, together with allowance for changes in stocks.

In the Mexican case the only reasonably reliable figures are those for imports and exports; there are virtually none for stock changes, and regarding production there exist tremendous contradictions between the decennial data of the farm censuses and the annual data of the Department of Agriculture. However, since census data are unsuitable for establishing consumption trends because they refer to single years which may be unrepresentative, the only option is to use the annual data with all their imperfections. (For instance, in several important commodities the historical series of per-acre yields show sudden upward revisions in a particular year, not resulting from any technological change, but merely from a revision in the method of data collection.)

The data presented in the following paragraphs refer to "apparent consumption" only, inasmuch as no allowances can be made for year-to-year changes in stocks.* However, by using five-year averages, the distortion arising from this omission should be largely eliminated. In making per capita

---

*The figures include consumption for all purposes including human and industrial use, seed, and livestock feed.

calculations the 1979 revisions of the 1960 and 1970 population census data have been employed, which have increased the previously published figures for 1960 by just over one million and those for 1970 by over two million (and the preliminary figures for 1980). To avoid over-elaboration we have concentrated on two trend periods: the first extending for twenty years from 1940–44 to 1960–64 and the second for fifteen years from 1960–64 to 1975–79 (Table 2.1).

TABLE 2.1

**APPARENT CONSUMPTION OF FARM PRODUCTS:
1940 TO 1980**

Commodity	Pounds per Capita			Annual Growth Rate	
	1940–44	1960–64	1975–79	1940–44 to 1960–64	1960–64 to 1975–79
Corn	339.1	379.1	406.1	0.6	0.4
Wheat	65.9	85.1	109.6	1.3	1.7
Rice, milled	7.0	11.0	11.7	2.3	0.4
Beans	21.4	39.7	54.6	3.1	2.1
Sugar	41.2	62.1	88.6	2.1	2.4
Sesame seed	5.8	8.6	4.7	2.0	− 4.0
Cottonseed	16.8	49.4	19.9	5.5	− 5.9
Copra	5.9	10.1	5.6	2.8	− 3.9
Peanuts	3.5	4.5	4.8	1.3	0.4
Safflower	0.0	2.4	20.0		16.2
Soybeans	0.0	2.5	27.4		18.4
All oilseeds as oil	10.1	21.1	27.5	3.7	1.8
Cotton, lint	8.9	8.4	5.8	− 0.3	− 2.4
Henequen	4.1	4.1	2.0	0.0	− 4.7
Coffee	2.6	2.6	3.7	0.0	2.4
Tobacco	2.6	3.8	1.7	2.0	− 5.3
Alfalfa	197.8	276.6	515.3	1.7	4.2
Sorghum	0.0	22.6	176.5		14.7
Potatoes	11.5	20.6	25.4	3.0	1.4
Tomatoes	9.1	16.5	27.5	3.0	3.4
Green chiles	2.8	5.8	13.2	3.7	5.8
Oranges	26.0	44.8	75.1	2.8	3.5
Bananas	29.5	43.2	49.2	1.9	0.7
Apples	4.4	5.6	8.5	1.3	2.8
Avocados	5.9	6.7	10.9	0.6	3.3
Peaches	4.5	3.9	7.1	− 0.6	4.1
Mangoes	9.3	10.4	16.1	0.6	2.9

TABLE 2.1
(concluded)

Commodity	Pounds per Capita			Annual Growth Rate	
	1939 –40	1959 –60	1979 –80	1939 –40 to 1959 –60	1959 –60 to 1979 –80
All meat	63.1	59.5	68.4	– 0.3	0.7
	Quarts per Capita				
Fluid milk	90.6	107.9	113.0	0.9	0.3
	Number per Capita				
Eggs	74	86	156	0.8	4.1

SOURCES: CROP PRODUCTS: Department of Agriculture, Bureau of Agricultural Economics, *Consumos aparentes de productos agropecuarios, 1925–1976;* for later years, Annual Crop Reports of Department of Agriculture.
LIVESTOCK PRODUCTS: author's estimates; see Appendix.
POPULATION: Population censuses, including officially revised figures from 1960 onwards.

NOTE: "Apparent Consumption" includes feed, seed, waste, and industrial use. Corn and beans in 1940–44 adjusted upward; see Table. 1.1.

Considering first the consumption of corn, the staple cereal of the Mexican people, per capita use has been increasing very slowly—earlier at 0.6 percent per annum and latterly at only 0.4 percent. The growth, such as it is, has been accounted for in the seventies entirely by an increase in feed use, direct human consumption being already on its decline. In wheat, on the other hand, consumption has been expanding steadily at between 1.3 and 1.7 percent annually, indicating a gradual replacement of corn by wheat in the average diet, a phenomenon associated with the process of urbanization. Rice, at just over 11 pounds per person, is a much less important cereal, and per capita consumption has almost ceased to rise.

Dry beans *(frijol)* are, like corn, a symbol of the diet of low-income Mexicans. Unlike corn, however, bean consumption continues to rise. Between the early forties and the early sixties it almost doubled in per capita terms, and from then till the late seventies rose by another 37 percent, which in an era of rising incomes is difficult to believe. Beans are mainly cultivated by subsistence farmers, and hence they are a crop whose yields are especially hard to evaluate. It may be that part of the recorded increase merely reflects improved statistical coverage.

As to the Mexican appetite for sugar there can be less doubt. At more than 88 pounds per head, consumption is one of the highest in the world, partly due to the Mexicans' addiction to soft drinks and sweet buns. Per capita intake has maintained a steady growth of 2.1 to 2.4 percent per year over nearly four decades. (One of the side effects has been an alarming increase in the incidence of dental caries and of diabetes.)

Among the crops that produce vegetable oils, a series of profound changes has occurred. Forty years ago three crops—sesame seed, cottonseed, and copra—together provided 85 percent of the nation's vegetable oil requirements, but by 1980 they accounted for only 40 percent. Sesame production has been difficult to expand since it is chiefly cultivated on poor soils in areas of low rainfall; cottonseed has depended on fluctuations in the fortunes of cotton which reached its peak in the sixties; copra production also peaked in the sixties when the area devoted to coconut palms began to decline. During the sixties and seventies these declining crops were replaced by safflower and soybeans, each of which has enjoyed a phenomenal expansion on the irrigated lands. Between them they now account for over half the vegetable oil produced.

Per capita consumption of these oils, as a group, was increasing rapidly until the sixties at 3.7 percent per year, but subsequently increased more slowly at only 1.8 percent (including imports). Considering that industrial uses in soaps, paints, etc. have been expanding, this means that direct human consumption has been either stationary or falling slightly.

Cotton production, both for export and for the domestic market, expanded rapidly until the late sixties, exports taking 70 percent of the total output. Thereafter export volume declined while domestic demand turned increasingly to man-made fibers. Per capita consumption has been falling at an annual rate of 2.4 percent, and no recovery seems in sight. Henequen, grown principally in Yucatán, has suffered a similar experience. It has lost part of its export markets to sisal, while at home it has been displaced in many uses by paper and plastic wrapping materials. Since the early sixties per capita consumption has fallen at a rate of 4.7 percent per annum.

Another tropical export crop, coffee, has recorded a more prosperous development, with production expanding to meet both higher internal consumption and higher export demand. High world prices in the mid and late seventies stimulated extensive new plantings. Until the sixties per capita consumption had been static at a little over 2.5 pounds, but subsequently the coffee drinking habit mushroomed, and consumption per person grew at over 2 percent annually.

Tobacco has had quite a contrary history. Consumption per person was rising steadily until the end of the fifties; since then it has halved, a peculiar phenomenon since the publicity given to the health hazards of smoking has been less intense in Mexico than in many other countries. The growers have been saved through developing a useful export business.

Mention should be made of two significant feed crops widely grown in Mexico. Alfalfa, which requires irrigation, has expanded rapidly; production more than trebled between the late fifties and the late seventies. Grain sorghums, first introduced into Mexico in the fifties, have proved highly successful with an output of around one million tons in the mid-sixties rising to nearly five million at the end of the seventies. This crop has supported the fast-growing poultry industry and is used in hog fattening.

Finally we may turn to consider a few of the more important fruits and vegetables. In many of these crops there are marked contrasts between our two historical periods. Thus, from the forties to the sixties per capita consumption was rising rapidly (at 3 percent and more per year) in potatoes, tomatoes, and green chiles, and reasonably fast in oranges and bananas. In other fruits demand grew much more slowly. From the sixties onwards the Mexican public took to fruit and vegetables, so that per capita consumption of most kinds began to grow at rates of 2.5 to 4 percent per year. Only in bananas, and to a lesser extent potatoes, a certain market saturation became apparent. Outstanding growth was registered in peaches, oranges, and avocados. This diet diversification can be related to urbanization and to higher incomes.

When it comes to estimating trends in the consumption of livestock products, the official statistics provide no help at all. The census data appear to underestimate the output by something like 50 percent, while the annual reports of the Department of Agriculture register chaotic year to year fluctuations, many of which would be biologically impossible. For these reasons, the author had to piece together some estimates, making use of evidence scattered through various monographs and comparative material from other Latin American countries in similar stages of development. In respect to livestock numbers it seemed preferable to use the census figures up to 1970 and then those in the annual reports for subsequent years. As to the output of meat, milk, eggs, etc. and levels of national consumption, rates of growth of productivity were established which gave reasonable results in earlier decades both for consumption levels and for herd performance (see Appendix).

The estimates presented in Table 2.1 refer to the census years 1939–40, 1959–60, and (based on annual reports) 1979–80, and must be considered as nothing more than best approximations. It will be at once noticed that, with the exception of poultry, the livestock sector has been far less dynamic than crop production. Since livestock products are relatively expensive, one naturally expects low levels of consumption and slow rates of increase among a population where the incomes of the majority are still low. This is compounded by the physical features of Mexico that are generally hostile to the animal sector—semidesert grasslands of poor quality in the north, or humid tropics infested with pests and diseases in the south and the coastal plains.

According to our estimates, meat consumption per capita actually fell slightly between 1940 and 1960, and since then has risen only very slowly. Had there not occurred a phenomenal expansion of poultry meat production, total meat consumption per person would not have increased at all. The share of poultry meat in the total has risen from 10 to over 20 percent during the forty-year period, an evolution accompanied by a sharp decrease in the relative price of this commodity.

Milk consumption has registered a trend hardly more encouraging. Up until 1960, it is true, there was some expansion, but in the following period this slowed down to about 0.3 per annum (although it was somewhat higher in the seventies than in the sixties). Milk is not widely used either in cooking or as a beverage except for small children. Furthermore, government regulation of milk prices has discouraged dairy farmers from increasing or improving their herds. National requirements have more and more been met by imports of powdered and condensed milk.

Only in eggs has consumption taken off in a big way, with per capita consumption more than doubling in twenty years. As a result of modern poultry technology, the real price of eggs to the consumer has been falling constantly, relative to the prices of other foods, and to this stimulus demand has proved highly responsive.

Among the minor livestock products, honey has recorded the fastest rate of growth, but output has been largely for the export market. Wool production, which once had some importance though its quality was medium to low, has declined partly because of lack of interest in sheep improvement and partly through the competition from man-made fibers. Hides and skins have never become profitable because of tick damage to the animals.

It is also useful to calculate what happened to the consumption of all agricultural products, taken as a group. This concept is derived by estimating the rate of growth of total agricultural supplies, measured at constant prices and adjusted for exports and imports of farm products. Thus during the two decades from 1940 to 1960, total supplies on the domestic market were increasing at about 4.5 percent per year, which with a 3.1 percent annual average growth in population, was equivalent to an annual increase of 1.4 percent per capita. In the two following decades the situation changed markedly. Population was expanding more rapidly at a rate of 3.5 percent, but since per capita income was rising more slowly than before, food consumption per person increased at only 0.9 percent. The combination of these two elements meant that the supply of farm products had to grow at 4.4 percent, and since at this same time (1960–80) the growth of agricultural production was only 3.8 percent, requirements could be met only by a substantial increase in food imports.

Some corroborative evidence on this subject can be found in figures published by the Food and Agriculture Organization of the United Nations, although they refer to a slightly different period (1952–72) and to food only. Citing certain Latin American countries, the FAO gave a per capita food consumption growth rate of 1 percent per annum for Brazil, Costa Rica, and Peru, 0.9 percent for Mexico, 0.7 percent for Venezuela, 0.5 percent for Colombia and 0.3 percent for Argentina. This suggests that Mexico was in the top group of countries in this respect; and bearing in mind that national averages are heavily influenced by changes taking place in the performance of the lower income groups that constitute the bulk of the population, it would appear that in Mexico the masses were improving their food consumption about as fast as any in Latin America during these years.[1]

The study of consumption trends and their use as a basis for consumption projections has accumulated a considerable literature in Mexico from the mid-sixties onwards. Like economic projections in other fields, most of these proved to be wide of the mark, and for this there were two principal reasons.

---

[1]Food and Agriculture Organization of the United Nations, "Population, Food Supply and Agricultural Development," background paper prepared by FAO for the World Population Conference, Bucharest, 1974.

The first mistake was the prediction that by the mid-seventies Mexico would face a serious problem of agricultural surpluses, whereas by then she was having to supplement insufficient domestic production with food imports. Here the fault lay mainly in assuming a linear continuation of the growth of agricultural output that had been registered in the years 1940–65. As was noted in Chapter 1, it was precisely after 1965 that the growth of output slackened, and it has not subsequently even kept pace with the growth in population.

The second error was the overestimation of the future volume of consumption of livestock products and underestimation of crop products. This arose from neglecting the price factor. Projectionists habitually cover themselves with some phrase about "other things being equal," but when one of the "other things" happens to be an important determinant of consumer behavior such an omission may be dangerous. In this case they failed to foresee that the retail prices of animal products would be rising more rapidly than those of crop products, partly because in a hostile physical environment the costs of livestock production were rising as output expanded and partly because the government was engaged upon a policy of subsidizing certain basic commodities, notably corn and sugar.

Most of these were eminent authors making serious contributions to analyzing Mexico's economic problems. However, it would not be chivalrous to harp on their mistakes. In what follows we shall probably stumble into different but just as serious pitfalls, and be easily shown wrong by writers ten years hence. Nevertheless, it is desirable to make some sort of demand projections in order to establish some quantification of the task facing Mexico's farm sector.

INCOME ELASTICITIES OF DEMAND

One of the most important tools used in demand analysis is the concept of income elasticity of consumer demand.

In very poor countries people tend to devote more than half of any income increase to food expenditure; in rich countries they devote one-third, or less. As regards particular foodstuffs, a rise in income tends to be associated with a very small or zero increase in expenditure on basic foods, such as bread and potatoes, and a much larger increase in purchases of meat and fruit, these having high elasticities. If, in the case of Mexico, we were equipped with a reliable set of income elasticities covering all the principal commodities, then, after making an assumption concerning the expected increase in disposable per capita income over some future period, we could go a long way toward predicting how the pattern of expenditure on those commodities would change. It is necessary to emphasize the phrase "go a long way toward" because, although income elasticities are important, they are not the sole determinant of expenditure, as will be seen presently.

Income elasticities of demand can be calculated by either of two methods: first, using data provided in surveys of family income and expenditure, and secondly studying changes in income and expenditure through time at the national level. In Mexico two major surveys of family consumption have been undertaken under the sponsorship of the Bank of Mexico, one in 1963 and the other in 1968. Not only are these rather out-of-date, but their methodology and reliability have been much criticized. Certainly the income elasticities which emerged in respect of particular foodstuffs were too divergent to be any guide at all, as Table 2.2 shows. Despite the fact that only five years separated the two surveys, we find differences of threefold and more in the coefficient for the same commodity.

TABLE 2.2

INCOME ELASTICITIES OF DEMAND
FROM HOUSEHOLD CONSUMPTION SURVEYS

	Urban		Rural	
	1963	**1968**	**1963**	**1968**
Corn	− .484	− .142	− .294	− .096
Beans	− .282	− .096	− .237	− .041
Beef	.222	.512	.927	.961
Beer	.515	2.289	.670	2.157

SOURCE: Banco de México, S.A., *La distribución del ingreso en México*, Fondo de Cultura Económica, México, 1974.

Whether or not there were defects in the execution of these surveys, there is a more fundamental objection to using this type of evidence. A family consumption survey is carried out at a single point in time, and when data are collected from a high income family it is probable that this family has already belonged to a high income group for a considerable number of years; likewise the low income families have probably been poor for as long as they can remember. Therefore, what is being measured is the consumption behavior of two or more disparate social groups, not what happens when a family moves from one group to the next. Such surveys may illuminate aspects of the contemporary structure of society but they cannot pretend to measure *change*.

Projections exercises are essentially concerned with change, with what modifications in consumption levels are likely to occur over a period of years as a result of assumed changes in incomes and in other factors. We are interested in how family A may alter its consumption when it moves from, say, the tenth to the ninth decile of income distribution, or family B moving from the fourth to the third. To use evidence from household surveys in this context would involve assuming that family A, after moving up, immediately behaves like those who have been in the ninth decile for years, and similarly with family B's behavior. But common sense says this is not so. The nouveaux riches do not behave like the aristocracy, neither does the bricklayer behave like a professional man as soon as his son qualifies in medicine. Household surveys give no clues as to what happens in these situations, which are the ones relevant to the present purpose.

FOOD HABITS

There is nothing more conservative in human beings than their attitudes toward food. They like what they are accustomed to and will continue to purchase simple foods long after they have money enough to purchase more expensive articles. In other words there is a time lag in adjusting to a higher level of income, one which may be as prolonged as a whole generation.

In Mexico, particularly in the cities, we observe many families who will devote an income increment to buying a used car or to moving into a better dwelling, while retaining their former dietary pattern virtually unchanged. They will continue to eat tortillas, tacos, and tamales. They will not increase their milk consumption, except powdered milk for the infants, partly because in most parts of Mexico milk does not keep unless one possesses a refrigerator. Furthermore, the very poorest people, who did not consume even a sufficiency of tortillas, will at last satisfy their hunger by buying more of these. This analysis helps to explain why, in spite of rising incomes in Mexico, the consumption of fruit and livestock products has not risen as much as was projected, while that of corn and beans has not declined.

These are some of the reasons why, for purposes of demand projections, it is safer to base our calculations on the growth trends historically observed, because these trends actually incorporate the phenomenon of conservative food habits in different strata of the population, the migrations from rural to urban areas and other influences of a social character. Such a choice by no means binds us to the method of linear projection; indeed in several important commodities a linear approach would not be practicable inasmuch as they have demonstrated very distinct consumption trends from one decade to the next. Comparing the sixties and seventies with the forties and fifties, we have noted very different rates of growth of consumption in such articles as rice, beans, coffee, and tobacco.

The real task is to try to isolate the factors which have caused these important shifts in the slopes of the growth curves, and then decide which of these several factors are likely to persist in the coming years. Historical series can be a serviceable guide only when they have been adjusted in the light of knowledge of their determinants. Some of these can be evaluated only in a qualitative manner, for example the effect of pronouncements by the medical profession concerning the health dangers of certain commodities. Other determinants may be quantitative and susceptible to mathematical treatment, and here among the most important are the prices of individual foods, and the changes, both absolute and relative, in these prices. (Another factor sometimes cited as significant is that of change in the pattern of income distribution, for instance toward greater equality; but experience teaches that such changes really occur very slowly—especially if we omit changes in social income—and would not affect our ten-year period. A much more relevant aspect of redistributive policy is the factor of food subsidies, which anyway will show up in our discussion of prices.)

PRICE ELASTICITIES
AND RELATIVE PRICES

Earlier it was noted how one of the shortcomings of Mexican authors' projections of consumption was the failure to take sufficient account of possible changes in the prices of individual commodities. In discussions of price as a determinant of consumption, two distinct concepts are involved. One is that of "the price elasticity of demand," which is formulated by calculating what percentage change will occur in expenditure on commodity A when its price rises (or falls) by one percent. The literature on this topic in Mexico is practically nonexistent, but it is generally assumed that in this country, as elsewhere, price elasticities of demand are greater for luxury foods than for basic articles of diet, and greater in lower than in higher income groups.

This in itself does not carry us very far. It is therefore necessary to focus on the second concept, namely the movement in relative prices—in other words how the price of A changes over time in relation to the price of B, and what influence this change has on the consumption of A and of B. For instance, if as a result of rapid progress in agricultural technology and a consequent reduction in production costs, the prices of all foodstuffs declined in relation to the prices of other consumption goods, this should provide a general stimulus to food consumption. The more likely case, however, is a fall in the production costs of certain articles, causing a shift in the *relative* prices of foods among themselves. A similar situation may arise when a government decides to subsidize certain foods and not others. Such changes in relative prices have more impact when they occur between two foodstuffs more or less substitutable, e.g., two kinds of vegetables or two different fruits.

In Mexico in recent decades important changes have occurred in the prices of food as a whole, as well as in the relative prices of individual commodities, and some of these have had a significant impact on trends in consumption. Considering first the movement of food prices as a whole, we find that food prices at wholesale level have increased just slightly faster than prices in general, but the difference is rather unimportant. At retail level the difference is considerable, with food prices rising significantly less than those of other consumer goods, the main explanation being the large subsidies provided by the government in order to diminish the effects of inflation on basic articles of diet.

Still more interesting and relevant for the present analysis are the movements in prices of individual groups of foods, which however and unfortunately are published only at wholesale level. The accompanying graph (Fig. 2.1) shows that the price of flour products (composed mainly of corn and wheat products) has remained for many years well below the general food price index (influence of the subsidies). The price index of sugar and sugar products has likewise trailed behind, due to the same influence (sugar subsidies being particularly high until the summer of 1980 when the price was doubled). In sharp contrast to these items has been the trend in the index of the livestock products group, which has remained consistently 25 to 30 percent

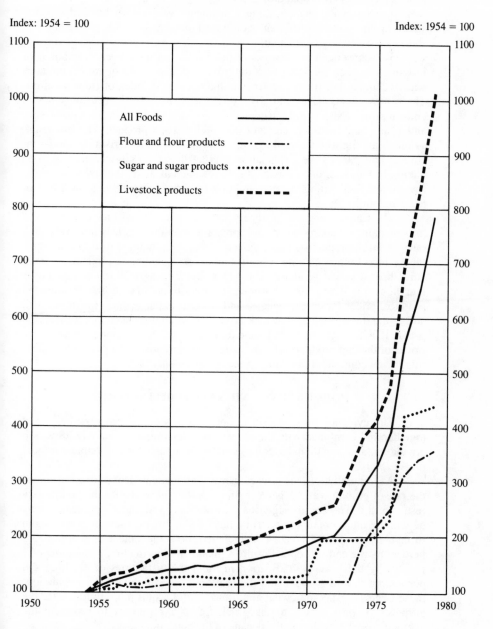

FIGURE 2.1

**WHOLESALE PRICE INDICES FOR SELECTED FOOD GROUPS:
1954 to 1979 (1954 = 100)**

Index: 1954 = 100

Index: 1954 = 100

All Foods

Flour and flour products

Sugar and sugar products

Livestock products

Source: Banco de México, S.A., annual reports.

above the general index. Although the authorities operated controlled prices for meat, milk, and eggs, they were obliged to raise them from time to time to reflect increases in production costs, notably of feeds. Even so, as will be seen in a subsequent chapter, these adjustments were usually too tardy and too small to stimulate a sufficient volume of production with the result that shortages frequently developed.

The consequential changes in price relatives have been dramatic. For instance, between 1954 and 1958 the rise in the price index of animal products was two and a half times greater than the rise in the index of flour products. These movements have naturally exerted a powerful influence on trends in consumption. Earlier in this chapter we noted how the consumption of corn and beans had not declined, and that of livestock products had not risen, as much as might have been expected during a period of rising personal incomes; and part of the explanation was attributed to conservatism in food habits. But another determinant is now revealed as being the changes in relative prices, stimulating the consumption of cereals and holding back that of meat and milk.

The future prospects in regard to price relatives will have to be taken into account in making our projections of demand to 1990. If relatively high rates of inflation persist, future governments will continue trying to mitigate their effects by maintaining subsidies on the so-called basic products, namely corn, beans, wheat products, and sugar. Because agricultural production is lagging and will need further stimuli, they are likely to authorize increases in producer prices for cereals, beans, and oilseeds in particular. This in turn will add to the difficulties of livestock farmers who do not depend exclusively on grass. It will augment their production costs and also, sooner or later, the prices of the end products. Hence the divergent trends of the past two decades might well continue, but relative price is not the only factor.

POPULATION AND NATIONAL INCOME

The evidence presented in the preceding sections can now be pieced together and combined with the other major determinants, namely growth in population and in national income, to provide a basis for our demand projections exercise.

As to population, the revised census data indicate that during the sixties the rate of growth was 3.5 percent. In respect to the seventies the preliminary results of the 1980 census suggest a growth rate of just under 3 percent. It may be that the family planning program launched in the mid-seventies has begun to take effect. The official target included the achievement of a rate of 2.5 percent by the end of López Portillo's presidential term (1982) and a rate of 2 percent by the year 1990. Bearing in mind the difficulties of reaching and persuading the urban poor and the scattered, conservative rural population, these targets appear somewhat ambitious, and it seems more prudent, for the purposes of projections, to take a figure of 2.7 percent to represent the average growth rate during the decade of the eighties.

As to national income, the projection problems are more numerous and difficult. Even in sophisticated and statistically well-equipped countries the econometricians have difficulties in forecasting one or two years, much less a whole decade. In Mexico, besides the usual perplexities, there are two additional major uncertainties. The first concerns the oil resources of the country, how they will be administered and what effects they will have on the rest of the economy. Possibly oil production will be kept at moderate levels in order not to cause too much disruption and inflation, and to permit further industrialization across a broad front. On the other hand, with high prices for petroleum prevailing in world markets, the temptation will be strong to seek maximum advantage from this situation, which might not endure many years. In this latter case national income would indeed rise rapidly, but would tend to concentrate in rather few hands and would do little to augment the purchasing power of the masses. Moreover, public investment would also be sharply increased. From the viewpoint of demand projections this would be equivalent to a rather slow growth in average disposable incomes.

Another factor of uncertainty which has come to play an important role in Mexico is the complexion of the government in a particular presidential term. So long as government intervention in the economy was minimal, a complete change of political leadership every six years had no startling consequences. But from the beginning of the seventies intervention increased rapidly, the public sector grew faster than the private, government spending far exceeded revenue, and inflation became endemic. During the later years of President Echeverría's term, the combination of these factors together with political and social instability resulted in a much slower growth of national product.

President López Portillo restored confidence, regained a high growth rate of GNP, but failed to reduce the inflation rate. Although he encouraged the private sector, except in farming, the public sector continued to become more dominant. During the eighties two new presidents will take office and much will depend on the policies which they choose to pursue. Thus, they might decide to cut down the budget deficits, reduce progressively the losses of the State enterprises and strengthen the private sector. Alternatively, they might pursue policies of further nationalization, of liberal social spending, and of greater intervention in the market economy. The consequences for the growth of national income could be very different in these two scenarios.

In view of these various uncertainties and others which might be added to the list, it seems prudent to adopt the technique of presenting two alternative projections of the rate of growth of national output, leading to two alternative demand projections. The first, which might be called the high or optimistic rate, sets the annual average rate of growth of gross domestic product at 8 percent during the 1980s. Admittedly this equals the performance of the best years of the fifties and sixties, but with the new factor of petroleum it should be realizable. The second or low rate we set at 6 percent, which might result if inflation became worse and if the private sector were submitted to discouraging restrictions.

At this point it is necessary to refine our concepts. What matters for demand projections is not so much the trend of gross domestic product but rather the rate of growth of private consumption expenditure—in other words what is left over after deducting government consumption and savings both public and private. For some time past the volume of government consumption and investment has been on the increase, and this trend must be expected to continue. It follows that private consumption expenditure in the eighties must be expected to grow somewhat less rapidly than gross domestic product. For this item we have arbitrarily selected growth rates of 7.6 and 5.6 percent for the high and low alternatives respectively (Table 2.3).

When we recall our assumption that population will be expanding at a rate of 2.7 percent per annum, we can conclude that private consumption expenditure per capita would grow at either 4.9 or 2.9 percent annually. The lower of these rates would approximate the average performance in other third world countries, while the higher would by any yardstick be outstanding.

FUTURE DOMESTIC DEMAND FOR FARM PRODUCTS

Armed with these assumptions and projections regarding the more general factors, we can now consider the consequences in terms of future trends in consumer demand. We will first examine the outlook for farm products as a whole and then that for individual commodities or groups of commodities. In both cases the initial step will be to determine appropriate coefficients for future income elasticities of demand.

In choosing these coefficients it seems preferable to be guided by the historical behavior of demand at the national level for reasons given earlier, rather than by the evidence of the household surveys. The historical data indicate that, over the past few decades, the income elasticity of demand for farm products as a whole has been around 0.5 with little change until the late seventies when there were suggestions of a slight fall.

TABLE 2.3
ALTERNATIVE PROJECTIONS OF CONSUMPTION
OF FARM PRODUCTS: 1980 TO 1990

	Annual Rate of Growth	
	High Alternative	Low Alternative
Gross domestic product	8.0	6.0
Private consumption expenditure	7.6	5.6
Population	2.7	2.7
Private consumption expenditure:		
per capita	4.9	2.9
Income elasticity of demand	.45	.5
Consumption of farm products:		
per capita	2.2	1.45
total	4.9	4.15

In our high alternative national output will be rising rapidly, and one must assume that the masses will participate to a substantial extent in this prosperity. Given a steady rise in incomes, the income elasticity of demand will almost certainly diminish; and to reflect this we have chosen a coefficient of 0.45. This means that with high growth the per capita consumption of farm products would increase at a rate of 2.2 percent (0.45 x 4.9) and total consumption at 4.9 percent. The details of the two models are set out in Table 2.3.

Since our low growth rate in consumer expenditure approximates that achieved in recent years, and since one should take account of conservatism in food habits, it seems justifiable to retain a coefficient of 0.5 to represent income elasticity of demand. This implies that during the eighties the per capita demand for farm products as a whole can be expected to grow at 1.45 percent annually, and total demand (including population growth) at 4.15 percent.

From this broad picture of trends in global demand for farm products, we can proceed to a consideration of individual commodities. Here the doubts and the possibilities of error are much greater, partly because the individual elasticities have not remained consistent through time and partly because it is especially difficult to predict changes in price relatives since these can be influenced by arbitrary decisions of government. However, if some of the individual commodity forecasts turn out to be a little too high and others too low, the magnitude of the overall challenge to the farm sector's production potential will not be seriously altered.

Basically it is here assumed that in circumstances of a slow rise in incomes the consumption of the staple foods, such as corn, wheat, rice, and beans will continue to expand, whereas with rapidly increasing incomes this consumption will expand very slowly or even fall in per capita terms. When this reasoning is applied to the several more important products the resulting chosen growth rates are those presented in Table 2.4.

Per capita corn consumption, it will be noted, is scheduled to rise slowly on the low alternative and to stabilize on the high one. Wheat, rice, and beans consumption grows faster in the first than in the second case. The only exception among staple foods is sugar for which demand has shown as yet no signs of saturation and is certainly consumed more widely as incomes improve. Hence for sugar, as for most of the remaining products, it has been assumed that higher incomes will be associated with higher consumption, bearing in mind always that it is the behavior of the masses which determines the trend in the national average.

A notable exception must be made for cotton because, as prosperity spreads, the shift to man-made fibers becomes more pronounced and demand for cotton declines. It appears probable that henequen and potatoes likewise fall into this category. Otherwise, for both vegetables and fruits we should expect a marked difference between the two alternatives.

With livestock products the principle is clear, namely that with enhanced purchasing power people will consume larger quantities, but the practical application is perplexing. Costs of production have been rising in recent years because of the problems with feed supplies, and this seems likely to

TABLE 2.4

ALTERNATIVE PROJECTIONS OF
CONSUMPTION OF INDIVIDUAL FARM PRODUCTS:
1980 TO 1990

	Annual Rate of Growth			
	High Alternative		Low Alternative	
	per capita	total	per capita	total
Annual crops:				
Corn	0.0	2.7	0.5	3.2
Wheat	1.3	4.0	2.3	5.0
Rice, milled	0.3	3.0	0.5	3.2
Beans	1.3	4.0	1.8	4.5
Sugar	2.8	5.5	1.8	4.5
Oilseed crops	1.8	4.5	0.8	3.5
Cotton	− 1.2	1.5	− 0.2	2.5
Alfalfa	3.3	6.0	2.8	5.5
Sorghum	10.3	13.0	8.3	11.0
Other annuals	2.8	5.5	0.8	3.5
Perennial crops:				
Henequen	− 1.7	1.0	− 3.7	− 1.0
Coffee	2.8	5.5	2.3	5.0
Citrus	3.3	6.0	2.8	5.5
Bananas	0.8	3.5	0.8	3.5
Other fruits	4.3	7.0	3.3	6.0
Other perennials	2.8	5.5	2.3	5.0
Livestock products:				
Meat	2.3	5.0	1.3	4.0
Milk	1.3	4.0	0.8	3.5
Eggs	3.3	6.0	2.3	5.0

NOTE: Population projected to increase at 2.7 percent per year.

persist. The steadily rising relative prices of meat and milk have so far curbed any expansion of consumption among the lower income groups. To take account of this, we have set the growth rates in the low alternative at only fractionally higher than those recorded in the sixties and seventies. But it has also to be recognized that in absolute terms per capita consumption remains low, even by comparison with other Latin American countries. Therefore, there probably exists a potential for considerable expansion of demand in circumstances of a significant rise in real incomes. The growth rates chosen in the high alternative are considerably higher than in the past, both for meat and for milk. Only in respect to eggs does it seem impossible that the rapid recent growth can be maintained much longer, and so some slackening in demand has been postulated. However, problems on the production side, especially with feed supplies, may cause milk and meat prices to rise still higher in relation to other foods. This would restrict consumption and prevent the demand targets from being realized.

EXTERNAL AGRICULTURAL DEMAND

To complete the demand picture a few words must be written about the evolution of external demand, howsoever complex the topic may be. In an overall sense Mexico's agricultural exports may be marginal to the farm sector; in value terms they amounted to less than 10 percent of farm output in the late seventies. But certain classes of farmers depend heavily on export markets, for example, the cattle ranchers of the north, the tomato growers of Sinaloa, and the coffee producers of the south.

The complexity of the world market needs no emphasizing. There are a great number and variety of importing countries, each with its own market characteristics; there are many exporting countries in each product fiercely competing with one another in price and quality. In a few commodities the trade is regulated to some extent by intergovernmental agreements, but mostly the markets are subject to violent price fluctuations every few years.

Mexico's chief farm exports have been and are cotton, coffee, tomatoes, and cattle (in part live and in part as meat). Among these commodities the market with the most favorable long-term prospects is that for beef and live cattle, and here Mexico enjoys a major advantage by her propinquity to the United States where demand is likely to remain strong, apart from certain cyclical import restrictions. However, unless she becomes a massive importer of feed grains, it is doubtful whether Mexico has sufficient natural pasture and feed of her own to sustain an expansion of this trade at the same time as satisfying the expanding internal demand for meat.

As to coffee, Mexico benefitted from the high export prices prevailing in the late seventies. But apart from periodic weather damage (such as frosts in Brazil), the world outlook according to FAO is that the exporting countries will by 1990 be producing more than the importing countries want to buy.[2] In such circumstances the fact that only a certain proportion of Mexican coffee reaches export quality, and even then is not of the highest grade, may make the competition difficult for her. In best years she has exported up to half her crop, but in the future her coffee growers may have to depend more on the home market which still has growth potential.

Cotton exports have long since passed their peak. Falling prices at home and abroad have caused Mexican growers to turn to other crops in the irrigation districts. Again the FAO predictions are that third world countries (together with other cotton exporters) will be producing surpluses, and prices

---

[2]Food and Agriculture Organization of the United Nations, *Agriculture Toward 2000,* Rome, 1979, C 79/24.

will remain weak. Mexico will continue to export, but not in the volume she attained in the sixties. The domestic demand for cotton is not expected to be buoyant, and the advent of oil prosperity is likely to intensify the shift to man-made fibers.

The lucrative tomato export business from Sinaloa depends entirely on the United States market, and periodically encounters strong opposition from growers in Flórida and elsewhere. Provided that satisfactory market-sharing arrangements can be concluded, for instance by quotas in the months when the two sources clash, this trade should continue to expand—as also that in tomato paste and in other vegetables and fruits. There have been hopes of developing new markets overseas for some of Mexico's tropical fruits, and these markets could become important, provided that Mexico's quality standards are raised and enforced, and that more professional effort is devoted to marketing and transportation.

One general comment affecting all Mexican exports has to be added at this point. The rate of inflation (25 to 30 percent at the end of the seventies) seems likely to remain, substantially higher than the United States' rate during the eighties. At the same time, the authorities have been and probably will continue to be reluctant to adjust the rate of exchange to reflect the widening disparity in prices. If this situation persists during the eighties, with devaluations too little and too late, all exporters, including those of farm products, will find their competitiveness progressively eroded. Mexico would be drifting into being a one-product (oil) exporter, which would conflict with official declarations in favor of broadening the industrial base.

This consideration, coupled with the view held by some influential Mexicans that it would be preferable to seek self-sufficiency in basic foods even at the cost of curtailing farm exports, might generate policies that caused these exports to decline certainly in relative and perhaps also in absolute terms. On the other hand, if a future government took steps to encourage a modernization of the farm sector along lines which will be discussed in later chapters, the export prospects could become very attractive.

THE AGRICULTURAL IMPLICATIONS

Finally, before examining each of the various physical resources—land, water, and livestock—at Mexico's disposal for meeting the demands projected in this chapter, it seems useful to consider briefly the overall requirement of farm land. With such an estimate in our minds, we shall be better able to evaluate the adequacy of our subsequent findings in respect to the physical resource potentials still available for development.

Since this exercise must by its nature by very approximate, we shall start by arbitrarily eliminating any changes that might occur during the eighties in the volume of agricultural exports and of agricultural imports. That is to say, we shall assume that exports and imports in 1990 will constitute the same proportion of national supplies of farm products as they did in 1980. It follows that if demand is projected to increase at 4.1 or 4.9 percent per year, agricultural production should be expanding at the same pace during the decade. If production failed to reach these targets, then either food imports would have to increase to make up the deficiency or there would be scarcities in national markets with rising prices which, in turn, would choke back demand.

Now the required increase in output could come partly from improvements in per-acre yields and partly from sowing larger areas. The question of yields will be discussed in more detail in Chapter 6, but we may here anticipate by saying that an annual increase of 2 percent, taking good years with bad, would be a good performance for the eighties, since during the seventies the gain averaged substantially less. If this much progress could be accomplished, then the remainder would have to come from increases in harvested area at average annual rates of 2.1 percent or 2.9 percent according to the low or high demand projections, which amounts to 22 percent or 33 percent over the decade as a whole.

Thus the harvested area, which around 1980 probably attained just over 40 million acres, would need to rise to about 49 (or 53) million acres. Furthermore, under Mexican conditions, the harvested area usually reaches only 60 percent of the arable area, and this proportion is unlikely to change substantially. What the total arable area was in 1980 is not yet known, but applying that same percentage to the harvested area, it probably was in the region of 64.5 million acres. By 1990 it would need to reach 78.5 or, on the high alternative, more than 85 million acres.

In other words, to meet the expected increases in demand, without increasing the share of imports, something between 14 to 20 million additional acres of arable land would have to be found—and this is after allowing an optimistically estimated rise in per-acre yields. This evaluation has been sketched out in a preliminary way in this chapter, leaving till later the several complications and qualifications, in order to indicate how tremendous is the challenge which Mexican agriculture faces when confronted by a rapidly increasing population combined with the near certainty of rising standards of living.

3. Land Availabilities

This chapter focuses on how Mexico's crop area was expanded to meet her increasing food requirements, and on what may be the potential for extending this process in the coming years. This represents merely a first stage in our search for ways of expanding food production, and it is solely a quantitative stage concerned with numbers of acres, leaving the various ways of making the land more productive by means of irrigation, fertilizers, pesticides, and so on for later discussion. We are concerned here with what sorts of land are still available for cultivation, where they are located, and what technical problems may be encountered in bringing them into use.

This search for cultivable land is not an activity peculiar to Mexico; it is going on in almost all third world countries as nations struggle to produce sufficient food for their increasing populations. The task is not easy. Between 1958 and 1978 while world population increased by 48 percent the world's crop area increased by less than 4 percent. Some countries are more fortunately situated than others in the sense that they have, or have had, relatively large areas of good land which they could bring into use; and for a long time Mexico was one of the fortunates.

Moreover, some countries have a more favorable "man-land ratio" than others. For instance, a country which possesses say three acres of arable land for every inhabitant is much more likely to be able to feed its people than a country which has only half an acre, leaving aside extreme differences in land quality, climate, and technology. Thus, Switzerland has only 0.15 of an acre of arable land per person and imports over half her food, whereas France has 1.0 acre and is almost self-sufficient. The U.S.A. has 2.2 acres and is a large food exporter, though her per-acre yields are well below those of Western Europe. The U.S.S.R. has nearly 2.5 acres, India 0.6, and Japan only 0.12 acres; all three are food importers. Mexico occupies an intermediate position. Back in 1930 there were 2 acres of arable for every Mexican, but by 1980 the man-land ratio had fallen to 0.75 acres as a result of the demographic explosion coupled with gradual exhaustion of the possibilities of finding more land.

HOW THE CROP AREA EXPANDED

The clearing of woodland and jungle, coupled with the mobilization of desert areas where irrigation became available, has been the principal contributor to expanding the crop area in Mexico. The historical evidence for measuring this process lies in the agricultural censuses which present, every ten years since 1930, a picture of land use in certain broad categories. This evidence has to be evaluated with some care, however, because of certain defects in carrying out the censuses.

Table 3.1 shows the basic data on land use from each of the first five censuses which purport to measure the totality of "land in farms." It can at once be seen how erratically this total has fluctuated from decade to decade, though in reality the 1940 figures were probably too low, while those of 1960 were certainly too high (partly because in 1960 a great deal of State forest got included and partly due to faulty reporting [see Appendix]).

The area of forest and other nonagricultural land pertaining to farms has also fluctuated significantly without any particular reason, and in 1970 this area was smaller than on any previous occasion. The pastures, on the other hand, appear to have been increasing slowly, apart from aberrations in the 1940 and 1960 reporting. They account for two-thirds of all farm land in the north and northwest, and an even higher proportion in Sonora and Chihuahua. (For names of states see Figure 3.1.)

While the vagaries of the census reporting of land categories so far mentioned do not greatly interfere with our analyses in this and subsequent chapters, when it comes to arable land—the area devoted to crops—the anomalies are so large that, if accepted as published, the figures would be unusable. Either one must abandon the attempt to learn something from the past or one must try to amend some of the more glaring errors. Here the latter course was chosen, and the errors were identified as being mainly of two kinds, (1) overestimation of the areas of shifting cultivation in the south and southeast, and (2) in the 1960 census a country-wide overestimation related to the authorities' first attempt to employ electronic processing. The original and the adjusted data are shown by geographical regions in Table 3.1, a panorama extending from the first farm census of 1930 to the most recent of 1970.*

For the country as a whole, the arable area increased slowly between

---

*For purposes of economic and social statistics, the federal authorities divide the country into regions, but no two agencies use the same regional groupings. For the present study the thirty-two states have been grouped into eight regions as follows:
  Northwest: Baja California Norte, Baja California Sur, Sonora, Sinaloa, Nayarit
  North: Coahuila, Chihuahua, Durango
  Northeast: Nuevo León, Tamaulipas
  North-center: Aguascalientes, San Luis Potosí, Zacatecas
  West-center: Colima, Jalisco, Guanajuato, Querétaro, Michoacán, Guerrero
  Center: Federal District, Hidalgo, Tlaxcala, Puebla, Morelos, State of México
  Gulf-and-south: Veracruz, Oaxaca, Chiapas, Tabasco
  Peninsula: Campeche, Yucatán, Quintana Roo
Although from an agricultural viewpoint each of these regions is more or less homogeneous, nonetheless wide contrasts can be found within each region, due to the many mini-ecologies which characterize Mexico. At this writing 1980 data are not available.

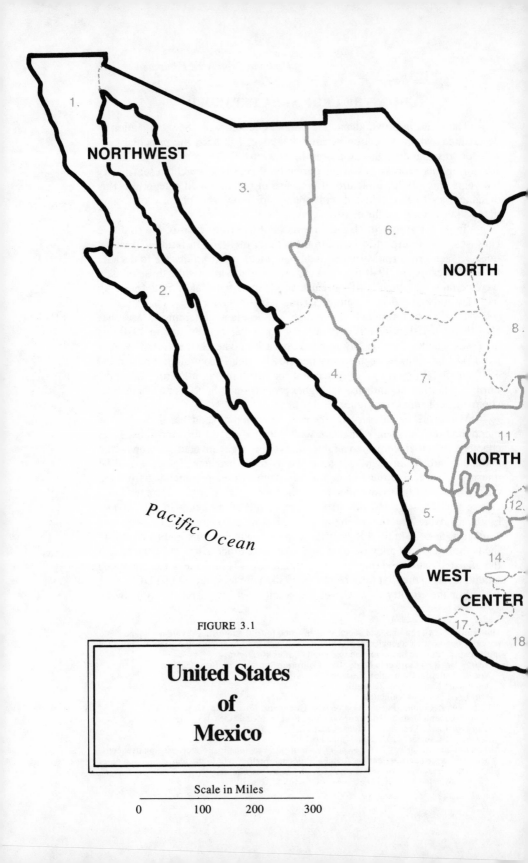

FIGURE 3.1

United States
of
Mexico

Scale in Miles

0	100	200	300

REGIONS
and
STATES

NORTHWEST
1. *Baja California Norte*
2. *Baja California Sur*
3. *Sonora*
4. *Sinaloa*
5. *Nayarit*

NORTH
6. *Chihuahua*
7. *Durango*
8. *Coahuila*

NORTHEAST
9. *Nuevo León*
10. *Tamaulipas*

NORTH CENTER
11. *Zacatecas*
12. *Aguascalientes*
13. *San Luis Potosí*

WEST CENTER
14. *Jalisco*
15. *Guanajuato*
16. *Querétaro*
17. *Colima*
18. *Michoacán*
19. *Guerrero*

CENTER
20. *Hidalgo*
21. *State of México*
22. *Federal District*
23. *Tlaxcala*
24. *Morelos*
25. *Puebla*

GULF-SOUTH
26. *Veracruz*
27. *Tabasco*
28. *Oaxaca*
29. *Chiapas*

PENINSULA
30. *Campeche*
31. *Yucatán*
32. *Quintana Roo*

NORTHEAST

CENTER

Gulf of Mexico

CENTER

PENINSULA

GULF-SOUTH

[45]

TABLE 3.1
LAND USE CHANGES, 1930 TO 1970,
AND REGIONAL DISTRIBUTION IN 1970

	Original Arable	Adjusted Arable	Grass	Forest	Other	Total Land in farms
	Mexico (in million acres)					
1930	36.1	32.4	164.3	64.0	60.8	325.2
1940	36.7	35.3	138.8	94.1	48.4	318.0
1950	49.2	43.4	166.5	95.9	47.9	359.5
1960	58.9	49.7	195.4	108.0	55.5	417.8
1970 unadjusted	57.2		184.0	49.1	55.3	345.6
1970 adjusted		47.3	193.9	49.1	55.3	345.6
	The Regions (million acres in 1970)					
Northwest	6.56	6.41	38.52	5.49	9.57	60.0
North	5.57	5.34	77.53	9.73	13.46	106.1
Northeast	3.45	2.98	15.44	5.76	3.31	27.5
North-center	4.51	4.10	17.22	2.26	5.71	29.3
West-center	12.04	11.64	20.37	6.02	7.32	45.3
Center	6.20	6.02	4.44	1.86	2.49	14.8
Gulf-south	16.57	9.23	17.66	7.43	8.15	42.5
Peninsula	2.27	1.56	2.80	10.51	5.26	20.1
	Percentage distribution in 1970					
Northwest	10.9	10.7	64.2	9.2	16.0	100.0
North	5.2	5.0	73.1	9.2	12.7	100.0
Northeast	12.5	10.8	56.2	20.9	12.0	100.0
North-center	15.4	14.0	58.8	7.7	19.5	100.0
West-center	26.5	25.6	44.9	13.3	16.1	100.0
Center	41.9	40.7	30.0	12.6	16.8	100.0
Gulf-south	39.0	21.7	41.6	17.5	19.2	100.0
Peninsula	11.3	7.8	13.9	52.3	26.1	100.0
Mexico	16.6	13.7	56.1	14.2	16.0	100.0

SOURCE: Statistical Office, decennial farm censuses.

NOTE: Total land in farms refers to the original (not the adjusted) arable areas, since it was not feasible before 1970 to distribute the adjustments among other land categories. For 1970 the only adjustment was to transfer to "grass" the 10 million acres of "cultivated pasture."

1930 and 1940, rapidly from 1940 to 1960, and actually declined between 1960 and 1970. Possibly a part of this decline is more apparent than real because our adjustments to the 1960 data may not have been severe enough; but some decline seems probable since another source (Department of Agriculture) shows declining harvested areas after 1966.

Regionally, the biggest increases between 1930 and 1970 occurred in the northwest, the west-center, and the Gulf-south, where in some individual states the arable area more than trebled (Table 3.2). By contrast in the states of the Altiplano, characterized by poor soils and low rainfall, the increases

TABLE 3.2

**INCREASES IN ADJUSTED ARABLE AREA BY REGION:
1930 TO 1970**

(in thousand acres)

	1930 – 50	1950 – 70	1930 – 70
Northwest	2,110	2,026	4,136
North	892	534	1,426
Northeast	1,450	425	1,875
North-center	1,369	109	1,478
Subtotal:	5,821	3,094	8,915
West-center	1,522	502	2,024
Center	549	146	695
Gulf-south	2,822	213	3,035
Peninsula	306	– 91	215
Subtotal:	5,199	770	5,969
Total:	11,020	3,864	14,884

SOURCE: Statistical Office, decennial farm censuses.

were in most cases less than 25 percent over the forty years; and these were mainly caused by demographic pressure. There are signs that, since the sixties, in some of these center regions the more marginal land is being abandoned: its inherent poverty coupled with soil erosion was producing very meager crops and miserable incomes.

In some quarters it is alleged that the big expansion of the cultivated area in the forties and fifties was due mainly to public investment in irrigation. The irrigation projects, among the finest investments which the Mexican government has ever made, certainly were decisive in the northwest where they accounted for most of the recorded increase, but over the country as a whole, though irrigation was important, it was not in quantitative terms the major factor. As can be seen from Table 3.3, from 1930 to 1970 the irrigated area rose by just under 5 million acres, while the rain-fed areas reached a peak of 41 million acres in 1960, a rise of nearly 13 million. The increase in rain-fed cultivation was widely distributed over most of the country, with 5.7 million acres in the four northern regions and 6.9 million in the four regions of the center and the south. The only region of minimal increase was the center for the reasons just mentioned. It is worth noting that the area of rain-fed cultivation more than doubled in the northwest and in the northeast, even though one is inclined to think of both as regions preeminently dependent on irrigation.

Furthermore, the expansion of the rain-fed areas started earlier than that of irrigation; some 70 percent of the rain-fed area increase was accomplished between 1930 and 1950, whereas the great bulk of the irrigation works was completed between 1940 and 1960. It may therefore be said that the first effect of an expanding market demand was to cause farmers to bring more rain-fed land into cultivation; somewhat later this effort by the farm people was aided

TABLE 3.3
AREAS OF IRRIGATED AND RAIN-FED ARABLE
BY REGION: 1930 TO 1970
(in thousand acres)

	1930	1940	1950	1960	1970
Northwest:					
Irrigated	734	954	1,663	2,716	3,274
Rain-fed	1,544	1,880	2,726	3,435	3,140
North:					
Irrigated	1,174	1,213	1,310	1,732	1,218
Rain-fed	2,745	3,039	3,501	4,023	4,129
Northeast:					
Irrigated	262	358	714	1,072	1,077
Rain-fed	845	1,297	1,843	2,039	1,903
North-center:					
Irrigated	124	166	188	292	306
Rain-fed	2,504	3,002	3,808	3,879	3,798
West-center:					
Irrigated	1,090	1,159	1,431	1,658	1,819
Rain-fed	8,531	8,681	9,711	10,509	9,824
Center:					
Irrigated	558	521	657	917	796
Rain-fed	4,764	4,710	5,214	5,140	5,221
Gulf-south:					
Irrigated	173	141	213	292	339
Rain-fed	6,019	6,837	8,802	10,185	8,888
Peninsula:					
Irrigated	30	2	12	7	22
Rain-fed	1,317	1,302	1,641	1,811	1,539
Mexico:					
Irrigated	4,144	4,517	6,187	8,686	8,854
Rain-fed	28,270	30,747	37,247	41,024	38,436

SOURCE: Statistical Office, decennial farm censuses.

NOTE: Irrigated areas as published; rain-fed areas calculated, by difference, from adjusted arable areas of each region.

and amplified by the government's irrigation programs. It was a period, as already emphasized in Chapter 1, when not only the technical but also the economic and social factors favored expansion: massive land distribution programs, falling production costs and favorable prices, highway construction programs which facilitated the transportation of farm produce, and prodigious successes in agricultural research.

What happened after 1960? According to the (adjusted) statistics, the arable area fell in ten years by nearly 2.5 million acres, all of it in the rain-fed category. Of this decline almost two-thirds were located in the Gulf-south and peninsula regions, which makes one suspicious that, perhaps, even after adjustment, there remains an element of exaggeration in the 1960 figures.

TABLE 3.4

UTILIZATION OF ARABLE AREA:
MEXICO, 1930 TO 1970; REGIONS, 1970

(in thousand acres)

	Arable Area (adjusted)	Utilization		
		Perennial crops	Annual crops	
Mexico:		area including fallow	gross sown area	gross harvested area
1930	32,415	1,806	30,609	15,921	12,856
1940	35,264	2,009	33,255	19,427	16,380
1950	43,435	2,024	41,411	24,818	21,285
1960	49,709	3,237	46,472	30,820	26,635
1970	47,290	3,583	43,707	30,215	26,172
Regions in 1970:					
Northwest	6,412	141	6,271	4,236	3,818
North	5,345	96	5,249	3,727	3,119
Northeast	2,980	208	2,772	2,080	1,735
North-center	4,104	158	3,946	2,873	2,260
West-center	11,643	467	11,176	7,260	6,135
Center	6,019	215	5,804	4,544	4,056
Gulf-south	9,226	1,583	7,643	4,943	4,564
Peninsula	1,562	716	845	553	485

SOURCE: Statistical Office, decennial farm censuses.

NOTE: Gross sown area includes the areas sown for second crops (mainly on irrigated land) during the year. Gross harvested area consists of the gross sown area less the preharvest losses. The 1970 figures of sown and harvested area were adjusted by the author for errors in the published data.

When travelling around the southern states one hears little of any large areas of land being abandoned during those years—and a 1.5 million acre decline is something which local people would notice. There occurred also some decline in the north-central and west-central regions, most probably of semiarid marginal lands. The other regions registered remarkably little change.

The arable area, from which cultivated grassland has been excluded, is devoted partly to perennial and partly to annual crops (Table 3.4). During the forty years from 1930 to 1970 the area under perennials shows the faster growth—an increase of almost 100 percent against only 43 percent for annual crops; and evidence from the Department of Agriculture's annual reports of harvested areas indicates that similar trends persisted during the seventies. This coincides with the finding in Chapter 2 that the demand for fruits is the fastest growing among all the groups of foods. Fruit and other plantation crops (coffee, palms, henequen, etc.) are located chiefly in the southern half of the country, particularly on the Gulf and Pacific coastal plains and (for henequen only) in Yucatán.

Table 3.4 also shows the area of annual crops harvested, as distinct from the area "dedicated" to these. The proportion of the dedicated area which is actually harvested has been increasing notably, from 42 percent in 1930 to 60 percent in 1970 which, as we shall see, reflects a reduction in fallow and in crop losses. One region, the center, is conspicuous for having always had a higher harvested percentage than any other—it was 70 percent at the last census—not because the center does not suffer from droughts, frosts, and other natural hazards, but because it is characterized by farms so small and so poor that their operators cannot afford not to harvest, however meager the quantities may be in a bad year.

INDIVIDUAL CROPS

Some quite remarkable changes have occurred during recent decades in the cropping pattern (Table 3.5). Between 1950 and 1980 (according to Department of Agriculture data) the harvested area of the so-called basic crops, that is cereals, beans, sugarcane, and oilseeds, has increased by 4.7 million acres, of alfalfa and grain sorghums by 3.6 million, and all other crops by 1.4 million.

Although corn and beans together increased somewhat in absolute terms, their share in the total harvested area fell from 71 to 54 percent. As the figures in the last two columns of Table 3.5 show, these two record the slowest rate of growth of any important crops except cotton, just as we saw in Chapter 2 that corn and beans were experiencing the slowest rate of growth of demand. Moreover, most of the expansion of corn and beans had already occurred by 1960. The wheat area also expanded rapidly during the 1950s, helped by the spread of irrigation and the development of the new high-yielding varieties, but during the past twenty years the area has actually declined. The other cereals, which include rice, barley, and oats, show a very modest increase in area, mainly due to an expansion of rice cultivation.

The two fodder crops listed separately in the table, namely alfalfa and grain sorghums, show an astonishing increase to over 3.7 million acres in 1978–80. Grain sorghums, although making their first appearance in Mexico as late as around 1960, have quickly become one of the most important crops in many parts of the country. To this huge increase in the area devoted to feed crops should be added the 10 million acres of "cultivated grasses," first recorded as a separate item in the 1970 census and which are also in the main an innovation of the last twenty years. Furthermore, the production of some 5 million of the acres devoted to corn and other cereals is destined for feeding animals rather than human beings directly.

In other words, the livestock population now requires for its maintenance, in addition to its permanent pastures and to considerable feed imports, some 19 million acres of arable land, which is nearly as much as the entire area dedicated to corn and beans. The significance of this large transformation

TABLE 3.5
HARVESTED AREAS OF SELECTED CROPS:
1950 TO 1980
(in thousand acres)

Crop	1949–51	1959–61	1969–71	1978–80 (prelim.)	Increase by 1978–80 (1949–51 = 100)	Annual growth rate (1949–51 to 1978–80) (%)
Corn	15,567	16,062	18,285	16,185	104.0	0.2
Beans	3,212	3,459	4,398	3,262	101.6	0.04
Wheat	1,522	2,155	1,927	1,680	110.4	0.3
Barley, Oats, Rice	988	1,260	1,087	1,186	120.0	0.6
Sorghum	0	279	2,256	3,212		
Alfalfa	132	225	383	534	404.5	5.2
Soybeans, Safflower	0	74	781	1,940		
Other Oilseeds	741	939	1,362	1,179	240.1	3.2
Sugarcane	479	830	1,280	1,196	249.7	3.3
Cotton	1,806	2,018	1,139	699	38.7	− 3.4
Cacao, Coffee, Henequen	790	1,205	1,455	1,483	187.7	2.3
Eight Fruits*	297	395	766	1,137	382.8	4.9
Other Crops	906	1,739	1,452	2,440	203.1	2.4
Total	26,440	30,640	36,571	36,133	136.7	1.1

SOURCE: Department of Agriculture, Bureau of Agricultural Economics, *Consumos aparentes de productos agropecuarios, 1925–1976;* and annual reports. Corn and bean acreage adjusted upward in 1949–51 and 1959–61.

NOTE: Data given for each of the four time periods (1949–51; 1959–61; 1969–71; 1978–80) are three-year averages.
*Apples, avocados, bananas, grapes, limes, mangoes, oranges, and peaches.

in the national cropping pattern will have to be examined in its livestock aspects in a subsequent chapter, but it is a change which has come to stay.

The oilseeds picture has been completely transformed by the arrival of soya and cártamo (safflower), which began to be cultivated in Mexico about the same time as sorghums and which now account for more than half the oilcrop area. Both are almost exclusively confined to irrigated districts, whereas the other oilseed crops are either like sesame, which is grown on non-irrigated land with quite moderate rainfall, or like oilpalms and coconut palms, which are located in the tropical coastal zones. While the soya and safflower areas are still expanding rapidly, the combined area of the other oilcrops has remained static for quite a long time. Cottonseed is of course also an important source of vegetable oil, but its production is determined by the demand for cotton, and the cotton area attained its maximum in Mexico in the late fifties.

The area cultivated with sugarcane increased rapidly and persistently until 1970, but since then there has occurred a decline which has eliminated the surplus which used to be available for export. The three principal tropical crops—coffee, cacao, and henequen—have more than doubled their combined area, the chief contributor to the expansion being coffee. However, together they occupy less than 5 percent of the total harvested area. The table also includes eight of the most important fruit crops whose combined area has expanded nearly fourfold since 1950.

In studying the past achievements and future potentialities for expanding the crop area, we have to bear in mind the geographical features with which the individual crops are associated. Thus, while the tropical crops are inevitably located in the two coastal belts of the southern half of the country, and the water-hungry crops such as cotton, oilseeds, and wheat are confined to the irrigation districts, the two Mexican staples—corn and beans—are concentrated on the Altiplano, chiefly from Zacatecas and San Luis Potosí in the north down to Oaxaca in the south. Although these are among the crops with lowest value in terms of output per acre, they (plus barley) are virtually the only ones that will grow in this low rainfall area unless supplementary water can be made available.

Furthermore, it has long been an area of demographic pressure where farms have been fragmented until there are now typically only 3 to 10 acres per family. Such low value crops on such small units result in miserably low incomes for the cultivators. This extensive zone, containing half Mexico's farm population, has been aptly described as the "rural poverty belt."

In this zone there are ten states where corn and beans account for more than 75 percent of the harvested crop area. What is more, the dependence on these subsistence crops has been increasing: it was higher at the end of the seventies than it was early in the fifties. This is an astonishing and alarming state of affairs. It is astonishing because one might have expected that the existence of Mexico City, and to a lesser extent Monterrey, with its rapidly expanding demand for a more diversified assortment of foodstuffs, for instance fruit and vegetables, would have promoted a shift away from corn and beans. It is alarming because it signifies a persistence of poverty, with a large number of farm families living almost outside the market economy and subsisting on diets unfavorable to health and good nutrition.

The observations in the preceding paragraphs underline the danger of drawing too many conclusions from national averages. When we were commenting on Table 3.5 we noted that the proportion of corn and beans in the total harvested area had fallen from 71 to 54 percent over the past thirty years; but now we see that in ten states the proportion has actually increased, while in another seven it still remains in excess of two-thirds. In other words, what appears to be the national trend in fact applies to only a portion of the country. In effect, there are two distinct areas, one in which the capacity for diversification has been proved to be strong, and another in which there is little potentiality for change. This theme will recur again and again in later chapters.

Before proceeding further, we will take a brief tour of the eight regions into which we have divided Mexico, because each one has distinctive characteristics in respect to its cropping pattern. Beginning with the northwestern region where over thirty years the harvested area has more than doubled, we note a pronounced trend toward crop diversification in all the states except Nayarit where the pattern has changed little. Baja California Norte abandoned its near monoculture of cotton, while the other states greatly reduced their earlier dependence on corn and beans as a result of the arrival of large-scale irrigation. This diversification took the form of a lot more wheat cultivation in Baja California Norte; more wheat, cotton, soya, and safflower in Sonora; and more of all of these plus tomatoes in Sinaloa. Some of the irrigation districts have managed to diversify more widely than others, one of the most successful being the Valle del Fuerte (Sinaloa), partly because of its abundant supply of water. A year by year analysis of the changes in areas devoted to individual crops in Sonora and Sinaloa shows a remarkable degree of responsiveness to market influences.

The northern region, comprising Coahuila, Chihuahua, and Durango, presents a less flexible agricultural structure, in part because it is a region with a more rigorous climate, poorer soils, and much less irrigation. The harvested area has increased only 15 percent since 1950, less than in any of the other regions. Changes in cropping patterns have been rather limited. Declines in corn, barley, and cotton have been compensated by moderate increases in sorghums and certain oilseeds and an enormous increase in the cultivation of beans. This is not a trend toward high-value crops, but merely a substitution of one group of low-value crops by another group. One exception has been the expansion of peach and apple growing in certain climatically suitable districts, but only a limited number of farmers participate in this activity. The only other exception of importance has been an expansion in La Laguna (on the borders of Coahuila and Durango) of milk production based on the irrigated cultivation of alfalfa. Apart from these isolated cases, the farmers of this region have achieved considerably less economic progress than those of the northwest. Indeed, in the extreme desert areas several thousand farm families still depend for their livelihood on the gathering of wild woodland plants such as *candelilla, ixtle de lechuguilla, ixtle de palma*, etc.

The northeast region is less homogeneous than most others in its characteristics and its evolution. Nuevo León belongs in crop matters partly to the northern region and partly to Veracruz (for instance its citrus), whereas Tamaulipas embraces several district zones: the corn and sorghums of the irrigated north, the sugarcane of the center, and the citrus of the center and south. The harvested area of Nuevo León has declined by 20 percent in thirty years while that of Tamaulipas has doubled. Nuevo León drastically reduced its corn cultivation, whereas Tamaulipas quadrupled its corn and has planted a million acres of sorghums to replace the cotton driven out (by pests and water shortage) from the irrigation districts on the Rio Bravo. Tamaulipas has also expanded its sugarcane and safflower. Both have developed citrus and other fruits and vegetables, which in Nuevo León now represent over one third of the harvested area, a higher proportion than in any other state of the Republic. Both have marginal areas of agricultural poverty, but in the main their farm populations have better resources and a more profitable range of crops than those of the northern region.

The region comprising Zacatecas, San Luis Potosí and Aguascalientes presents something of a puzzle in its evolution. Aside from its few and very small areas of fruit and sugarcane production and its livestock fattening on the pastures of the Huastecas, this region is characterized by what is one of the most hostile ecological environments for agriculture in the whole country. Its rainfall is low and erratic, most of its soils are eroded and lacking in organic matter, and it suffers severely from frosts and from winds. In spite of all these drawbacks, it has expanded its harvested area (if the statistics are to be believed) more than any other region except the Gulf-south; and it has accomplished this not by any sort of diversification but by remaining conservatively resolute in the cultivation of corn and beans, which together still account for 93 percent of the region's crop area. It has more than doubled its corn area and more than trebled its beans. Average per-acre yields of both these crops were thirty years ago, and still are, the lowest in the country. It is a region little affected by any form of technological progress. Poverty is endemic. There is some permanent and much temporary emigration, but the rate of natural increase remains high.

What we have called the west-central region, which stretches from Colima to Guerrero and inland to Querétaro, has almost doubled its harvested area and, in certain states, has experienced important changes in cropping patterns. The largest changes have occurred in Colima and Guanajuato and the least in Jalisco and Guerrero. Colima has greatly increased its orchard and plantation crops and thereby reduced its former dependence on corn and beans. Guanajuato has done the same, especially in the Alto Rio Lerma irrigation district, and is now cultivating a large area of sorghums as well as a wide range of vegetables and some fruit for the markets of Mexico City and Guadalajara. Querétaro has also exploited its proximity to the Federal District to expand its fruit and vegetables, but lacking irrigation this state remains heavily dependent on corn and beans.

Michoacán has had its ups and downs. For example, at one time its irrigated lands were producing cotton and other high-value crops at a good level of technical efficiency; whereas now most of the irrigation districts have fallen into neglect, cotton has declined notably from its level in the sixties, and the only significant increases have been in sorghums, peas, sesame and certain fruits such as melon. Perhaps the most surprising states are Jalisco and Guerrero, both of them large, with a combined harvested area which has doubled to five million acres, and yet their cropping patterns have remained virtually unchanged, both being heavily dependent on corn and beans. Otherwise Guerrero has its sesame in the upland areas and its palms along the coast. Jalisco's wheat and oilseeds have declined, to be replaced by sorghums and a little sugarcane. Of the two states Jalisco is the more diversified and has more pockets of agricultural prosperity.

The central region is one in which the harvested area has increased only slowly (little irrigation being available), and in which dependence on corn and beans has actually increased, this being particularly the case in the states of México, Hidalgo, and Tlaxcala. A sharp decline can be seen in barley, the typical cereal of upland areas with poor soils, while wheat has been almost obliterated. It is a region that grows hardly any sorghum except in Morelos and Puebla. Sugarcane has not increased at all, but there has been some useful, if limited, diversification into fruit and vegetables in the zones where natural conditions permitted, especially in Puebla. That this has not been more widespread in response to the enormous market offered by Mexico City can be explained only by the fact that most of the land in these states, with eroded soils, frost hazard, and erratic rainfall, is little suited to truck crops. Were it not for the many employment opportunities in the capital city, which enable thousands of families to continue occupying hopelessly uneconomic mini-farms, this region would suffer even greater agricultural poverty than it does.

The Gulf-south, the largest region of the eight, possesses some of the richest lands in the republic and also (in Oaxaca) some of the poorest. It has been expanding its agriculture more rapidly than any other region, but, considered as a single unit, its cropping pattern has not much altered, almost all of its traditional crops having increased in roughly the same proportion. Exceptions have been the remarkable expansion of sugarcane in Veracruz and of rice in Veracruz and Oaxaca. Both these states, together with Tabasco and Chiapas, have large areas under plantation crops which have been increasing continuously, and which embrace a wide range of fruits as well as coffee, cocoa, coconut and oil palms. Cotton has become significant in Chiapas but at a low level of per-acre yield. In all these four states the zones which possess .the best soils and the most favorable supplies of water are also exposed to other hazards such as cyclones, inundations, and an innumerable variety of crop pests and diseases, all of which combine to impede improvement in the yield and quality of the farm products. It is also surprising, in view of the abundance of water in most parts of the region, that as much as two-thirds of its area should continue to be devoted to corn and beans.

We come lastly to the peninsular states which, all three together, have a smaller harvested area than, say, Puebla. This area is still dominated by the henequen of Yucatán occupying more than one-third of the total. In both Yucatán and Quintana Roo the dependence on corn and beans has been increasing, whereas in Campeche some diversification into rice and sugarcane has commenced but encounters severe technical problems which will be referred to presently. Some fruit production for export was initiated with little success, since the main domestic markets are at too great a distance. The henequen industry faces inevitable decline due to the competition from man-made fibers; there is an urgent need to find some other crops as substitutes in that area, although none is as yet in sight.

From this brief journey through the various regions of the country the dominant impression one retains is that of immense contrasts in ecology, in cropping patterns, and in agricultural performance, contrasts which would become even more violent if one had space to describe the innumerable mini-regions, instead of merely the "macro" ones which are themselves conglomerates. There are indeed many Mexicos, and in respect to farming each has been evolving in its own distinctive manner. In certain regions and sub-regions a tremendous amount of change has taken place over the past thirty years, more especially in the northwest, the northeast, and the Bajío (Guanajuato) where, with the advantage of irrigation or good rainfall, new crops have been introduced accompanied by modern technology. Other regions have changed little, their farm families remaining dependent on the traditional corn and beans identified with deep and enduring poverty. It is this group which, in the years ahead, will have the most difficulty in adapting to the changing food requirements of the Mexican public.

PROSPECTS FOR MORE ARABLE LAND

From this brief review of the expansion of the arable area and of individual crops, we may now turn to the prospects for the future. It became clear in Chapter 2 that if the production goals required by the demand projections are to be met, a substantial contribution must be expected in terms of acres; not all of the production increases can be accomplished by improvements in per-acre productivity. So the questions must be posed: where can one find new land suitable for clearance? how much is available and at what pace could it be brought into cultivation?

These are difficult questions because very little pertinent information exists which might provide the bases for answers. There have been many vague statements in the agricultural literature that Mexico could cultivate an additional 20 to 25 million acres, but these statements have not been documented nor pinpointed to particular zones.

We know in general terms from meteorological information the distribution of the land area between arid zones and zones of good rainfall and, from aerial and ground surveys, the amount of land with a slope of less than 10 percent—any steeper land would require expensive soil conservation works (terracing, etc.) before it could be cultivated. Thus it has been estimated by Orive Alba that Mexico possesses some 173 million acres of relatively flat land with slopes of less than 10 percent, but of this amount only 31 million acres are in areas which could be cultivated without irrigation.[1] However, according to the 1970 census there were 38 million acres of non-irrigated land in cultivation; that is to say there were already 7 million acres located on lands too dry or too steep or both, i.e., the marginal lands. This is hardly an encouraging start to a search for new areas.

Because of this it may be prudent to begin with the negative features, by identifying the regions where we *cannot* expect to find more land.* Consider

---

*The present discussion is limited to the search for new land to be obtained by clearance. The potentialities for new areas of irrigation, and extending the crop area by that method, are postponed for consideration in the next chapter.

[1]Navarrete, Ifigenia M. de, ed., *Bienestar campesino y desarrollo económico*, Fondo de Cultura Económica, México, 1971, p. 99.

first the central region which has been inhabited and cultivated for several thousand years, and where in recent decades the multiplication of population has obliged farmers to cut down forests and push their ploughs up the mountain slopes in order to grow sufficient corn for their needs. In the state of México for instance, the general opinion of agronomists is that an excessive amount of land is being cultivated. In the northern part of the state one finds very shallow soils where the yields of corn and beans are extremely low. In the southern part, where corn has to be planted with a stick in order to retain the weed cover and thus check erosion, it is feasible to sow and harvest only every second year, and the maximum yield even with improved seed and with fertilizer would not exceed 800 pounds per acre. Some of these areas that are unsuited to annual crops could be converted to cultivated pastures; but in that case a farm family, instead of having the two, four, or six acres which it now operates, would need many acres to provide a satisfactory income. And what is found in the state of México can be found also in many other states of the central plateau.

If we turn to the northern regions, these, with the exception of parts of Tamaulipas, lie entirely in the arid zone, and those arable areas which do not enjoy irrigation are at the mercy of the insufficient and often erratic rainfall. In several years out of ten the rains may come so late that the sowings have to curtailed or, worse still, the seed sown germinates poorly, yielding only half a crop or no crop at all. In short, most of the land in these northern states is marginal in the sense of sustaining nothing better than a high-risk type of farming, which provides adequate harvests in only five or six years out of ten. This risk could be reduced in the few zones where additional irrigation could be introduced, but apart from this potentiality it would be folly to contemplate converting to crop use any of the pasture or remaining forest in these regions.

What then are the possibilities of the coastal zones—the "tierra caliente"? In principle these offer much more favorable conditions for farming. They enjoy higher rainfall and, at least in the plains, most of the land is not badly eroded. However, the rains tend to be badly distributed, with heavy concentrations in just a few months of the year, so that many districts may suffer from flood and from drought during the same season. Moreover, these areas have also been inhabited for centuries, and the lands most suitable for agriculture are already in cultivation. Nevertheless, along the Pacific coast from Nayarit to Chiapas one can observe several pockets of land which would be suited to clearance and cultivation, for instance along the Costa Chica of Guerrero and on into Oaxaca. In total they do not add up to hundreds of thousands of acres, but they would constitute a useful addition to the nation's resources of crop land. The fact that they are not being used is explained chiefly by the land tenure problems which will be analyzed later.

The greatest hopes for incorporating new land lie undoubtedly in the Gulf states from Tamaulipas to Quintana Roo. In Tamaulipas, especially the northern parts, considerable land clearance activity is currently to be seen, and the optimists speak of clearing 750,000 acres during the eighties, much of which would be planted with sorghums and some with cultivated grasses. In the Huastecas there remains a good deal of land on gentle slopes still covered with low forest or rough pasture which could be mobilized for crop growing. Much of this land is in ejidos whose officers have herds of cattle grazing these areas and who, therefore, have no interest in allowing them to be cultivated. Some of the land is owned by cattle ranchers who cannot cultivate it without changing the legal status of their farms under the Agrarian Reform Law.

Opportunities on a much vaster scale are believed to exist in the south and especially in the peninsula states; indeed, some people have gone so far as to claim that this region can and will solve Mexico's food supply problem. Already some ambitious projects are in various stages of completion: for instance, the Chontalpa project in Tabasco and the Edzná valley project in Campeche, and many similar projects are on the drawing board. So it is important to consider what their prospects are and what may be the scope for a still larger number of such enterprises.

Let us consider first the extremely mountainous state of Chiapas, where the indigenous (mostly non-Spanish speaking) tribes already cultivate slopes so steep that serious erosion results. It would be illusory to suppose that in this state anything more than a few pockets of land would be really suitable for clearance. In Tabasco the problem is different, because a high proportion of the area is in grass, and the proposal would therefore be to transform part of the pasture into cropland. But there is a good reason, as the local inhabitants well know, why so large an area has remained in pasture, namely that it is subject to flooding every year. The traditional system has been to pasture the cattle on the low-lying areas during the dry season and transfer them for the rainy months to the higher areas—not much higher, it is true, but a few feet in Tabasco make all the difference. To use these drier areas for crops would leave the animals nowhere to go, destroy the balance of the cattle industry, and produce less food than before. Much money has been spent on trying to control the floods by regulating the major rivers of the zone, until the engineers concluded that the cause was not the rivers but the rising water table during the wet season and that the only remedy would be a extremely costly chain of pumping stations along the coast to discharge the flood water into the sea.

In the three peninsula states the agronomic aspects are totally different from those of Chiapas and Tabasco. Some 80 percent of the area is classified in the census as unproductive, altogether nearly 16 million acres. This vast area of flat and virtually idle land (it produces some chicle) indeed constitutes a tempting prize. It is not surprising that certain people, desirous of creating for themselves a conspicuous political image, have talked of these 16 million acres as Mexico's future breadbasket. Yet this sort of pretension is primarily a matter of arithmetic. Somebody tells them that there remain in the peninsula 16 million acres of land not in agricultural use, and on the basis of this information they jump to their conclusions.

It would be preferable to analyze soberly the facts of the region, its climate and its agronomy. After all, the problem is not a new one in this particular peninsula of the great American continent. It has been facing the local inhabitants for more than two thousand years. The Mayas would appear to have achieved and maintained for many centuries an equilibrium between their food needs and the capabilities of their lands, and they found this on the basis of shifting cultivation whereby they took one crop every four or five years, or alternatively two successive crops followed by eight or ten years of resting the land. Toward the end of their civilization, peace and political stability caused the population to start increasing and therefore the number of crops taken every ten years had to be augmented. Consequently the land became exhausted, food supplies declined, and this contributed massively to the collapse of Mayan civilization.

Mexico is confronted with a similar demographic problem in the next few decades, and stands in mortal danger of making the same mistakes as the late Mayas. That this is not romantic theorizing can be confirmed by anyone who takes the trouble to travel over the peninsular states. For example, in northeast Yucatán attempts to improve the land by subsoiling have done the reverse, bringing nothing but stones to the surface. In the south of that state cropping is being attempted on impermeable clays which at first are waterlogged and then, after a few seasons, become unusable through progressive salination. In the Edzná valley in Campeche bulldozers have cleared the jungle for the first time in history, the exposed soils have rapidly become oxidized, the organic matter has declined and disappeared, crop pests and diseases have become rampant and uncontrollable.

The management of tropical soils for purposes of continuous cultivation over a long period of years, is a subject on which unfortunately very little is yet known. For a long time there were few agricultural research stations of any importance in the tropics. Now a certain number have been established and some of these have turned their attention to these tricky conservation problems. However, by the nature of the subject, no early answers can be expected, since alternative rotations and other techniques of soil management have to be tested over a period of several years (ten at the least) before valid conclusions can be drawn. Such research is inevitably location-specific. The ecologies of the Asian, African, and Latin American tropics are far from being identical; what succeeds in one tropical region may fail in another. That is why, if Mexico desires to develop agriculture in the peninsular region, it is essential that two or three well-staffed research stations be set up in contrasted zones to study this problem in all its aspects.

These comments and warnings should not be taken to imply that the peninsula of Yucatán has no agricultural future. It almost certainly has a favorable one, though not by using the conventional patterns of annual cropping. But while waiting for solutions to be found it will be prudent to proceed cautiously, not launching into grandiose projects master-minded by bulldozers. A priori it may be guessed that eventually the most successful technique may be some modernized variant of the system practiced by the Maya and other indigenous peoples. Instead of leaving the soil idle for a long resting period during which it reverted to jungle, we may be able, for instance, to plant green crops for animals, which would combine production with soil protection. But the basic principle would remain the same, namely adhering to a rotation of crops which satisfactorily maintains soil fertility and keeps the crops healthy.

If considerable space has been devoted to this discussion of the agronomic problems of the peninsula, it is fully justified by the tremendous importance of this topic for Mexico's future food supply. The peninsula possesses the last remaining untapped agricultural resource of considerable size in the whole republic. This resource may be exploited rashly and irresponsibly and to a large extent destroyed, or it may be harnessed skillfully so as to become a major supplier of food on a continuing basis.

But one should not expect too much. Those who claim that 16 million acres can be put under crops would do well to visit some of the costly and massive failures in Africa and elsewhere; studying the errors of other nations helps us to know what to avoid. It would probably be more realistic to expect that at a maximum some 10 million acres could be brought into agricultural use in that area before the end of the century, but of that 10 million not more than 2.5 million would be under annual crops in any one year.

Even 2.5 million acres would be a welcome addition to the sown area, and it will be worth waiting several years for the research results to become available in order to be sure of having these acres on a continuing basis. In addition to this resource there are the smaller areas already identified, mainly up the Gulf coast but also to a lesser extent on the Pacific side, which probably add up to more than a million acres and which could be brought into cultivation earlier, since most of them do not pose such difficult agronomic problems. Hence the overall conclusion of this section is that over the next fifteen to twenty years land clearance projects could probably provide an additional 3.7 million acres of arable land, plus an unquantifiable but important area of cultivated pasture.

FALLOW LAND AND PRE-HARVEST LOSSES

The amount of land effectively in crop production can be increased not only by bringing new land into cultivation but also by other methods. One of these is to increase the proportion of the existing arable land which is actually harvested, either by cutting down the fallow area or by reducing the proportion of sown crops which for various reasons is lost before harvest.

The amount of unsown (fallow) arable land at any one time is a function of ecology and ignorance. Thus there are certain types of soil which under particular climatic conditions need to be rested at regular intervals, perhaps once in four years, perhaps every second year, or in special cases as much as three years in four. A different example is the unexpected drought which prevents the farmer from undertaking sowings that year, a danger against which there is little defense except where possibilities still exist for introducing irrigation.

The techniques available for reducing the fallow area include: the terracing of sloping land which conserves the moisture from earlier rains, the sowing of green crops (between two annual crops) which can be ploughed in with a moisture-retaining effect, the inclusion from time to time of a nitrogen-fixing crop which improves soil fertility, and so on. Ignorance comes into the picture when farmers lack the know-how for adopting these various practices, when they persist with traditional soil-exhausting customs. Nonetheless, it must be recognized that in many parts of the Mexican Republic the ecologies are so unfavorable that even the best techniques could not eliminate the need for a considerable amount of fallowing. This is another way of saying that a great deal of extremely marginal land is being cultivated by people who have no alternative means of livelihood.

Preharvest crop losses result from droughts (after sowing as distinct from before), from floods, frosts, pests, and diseases. Post-sowing droughts can be mitigated only where some irrigation water can be obtained. Floods caused by cyclones or by an excessively high water table can be in part controlled by drainage projects, small-scale or large. Against frosts there is little defense except in orchards equipped with heating devices. In some local circumstances the remedies are too costly to be economic.

Severe crop losses may be caused by pests and diseases, or by birds or rodents. In some parts of the country the farmers have fields sufficiently large and have the equipment to apply modern control techniques, reducing their losses to a minimum—for instance, in the irrigation districts of the northwest. In other areas farmers lack the know-how and/or the credit to purchase the control equipment. The most difficult are the zones of mini-farms, not because of any difficulty in controlling a single acre—a knapsack spray-pump can be perfectly adequate—but because, if the cultivator of the neighboring plot does not also use pesticides, a farmer's efforts may be in vain.

In order to provide a basis for estimating the future potentialities of winning more cropland from reductions in fallow and in pre-harvest losses, it will be instructive to see what has been happening during recent decades. Let .us begin with the statistics of fallow.

Conceptually the fallow area is most easily calculated by removing the area under permanent orchard and plantation crops, and then calculating the fallow as the difference between the area under annual crops and the remaining arable area. Reckoned in this way the fallow percentage fell from 48 percent of the total arable area in 1930 to 35.8 percent in 1960. (Unfortunately defects in the 1970 census prevent us from using its data.) Fallow was always highest (60–75 percent of the total) in the south and southeast, partly due to shifting cultivation and partly to faulty enumeration; but it has declined to around 40 percent. It is lowest in the central region (24 percent), not because of any lesser agronomic need for fallow (probably the need is as high as, if not higher than, elsewhere); rather it reflects the intense demographic pressure which, as already mentioned, has obliged farmers in this zone to continue year after year cultivating lands which should really be resting periodically.

Altogether between 1930 and 1960 some five million acres were gained by reduction of fallow, but the gains were more rapid in the earlier than in the later part of this period, indicating that the possibilities were becoming reduced. In the states of the Altiplano there occurred hardly any further reduction after 1940. In subsequent decades the only notable improvements were associated either with the irrigation of the northwest or with a reduction in the practice of shifting cultivation in the south.

In the light of these trends and influences, what are the prospects for further gains over the next, say, ten years by means of further reductions in the fallow area? A good deal depends on the extent to which some of the much-fallowed marginal rain-fed lands can be given irrigation. There exist projects for major works in the northeast and the northwest, and there are potentialities for harnessing more groundwater resources in limited areas of the Altiplano and in certain zones of Oaxaca and Chiapas. The very ambitious irrigation plans for the south should not enter into the present calculation since they will be designed mainly to open up new areas, a prospect we have already taken into account. Concerning the fallow associated with shifting cultivation we should also note opposing trends. Population pressure in the indigenous areas is still impelling farmers to dangerous reductions in their land-resting practices, while in other areas the extension workers are increasing the awareness of conservation needs. These trends will probably cancel each other out.

In the absence of more detailed studies, it is difficult to put a figure on any projection for future economies in fallowing. Quite possibly the expected increases in irrigation in the northern regions could bring about economies of the order of 500,000 acres during the next 12 to 15 years. If we add to that another 250,000 acres to represent the net effect of technical progress of various kinds in other parts of the country, we reach a total of 750,000 acres, which is probably on the generous side. This is another way of saying that no substantial contribution to an expansion of sown areas can be counted on from this component in the medium-term future.

Let us now turn to the prospects for reducing the pre-harvest losses. During the thirty years from 1930 to 1960 (again the 1970 figures are unusable) more than 1.2 million acres were economized by progressive reduction in these losses, but again most of the gains were recorded in the earlier years. Such losses are traditionally highest in the low rainfall areas of the northern regions, reaching 20 percent of the sown area, suggesting that droughts (and sometimes frosts) are more decisive factors than inundations.

What under Mexican conditions are the potentialities of the available techniques for countering pre-harvest losses? Recognizing that virtually nothing can be done about cyclones and (over large areas) frosts, there remain the control possibilities against droughts, other inundations, pests and diseases. With regard to antidrought measures, in the four most vulnerable states, namely Coahuila, Chihuahua, Durango, and Zacatecas, there are few prospects of large new irrigation projects. Inundations affect chiefly the Gulf states from Tamaulipas to Campeche, and in these the progress in providing protection will be slow because costly. In the matter of pests and diseases, aside from the major irrigation districts, already well protected, the problems lie chiefly in the tropical coastal areas where in some years the losses are high. But these are the areas most difficult to control, because of the multiplicity of vectors and because of the little know-how and great poverty of most of the farm operators.

Again, as with fallowing, it may seem pretentious to try to quantify the influence which these various factors may have in the coming years. But we are dealing here not with prophecies but rather with identifying the elements which are likely to be of major or of minor significance. In this spirit, and bearing in mind that during the past twenty years or so the gains were probably less than 500,000 acres, we may hazard the guess that over the next 12 to 15 years the economies from reduction of pre-harvest losses are unlikely to exceed 250,000 acres.

DOUBLE CROPPING

Taking two crops in one year, or in some rotations three crops in two years, is another way of getting a larger area of crop from a given area of arable land. Most of this double cropping depends, in Mexico, on plentiful irrigation in order that both crops can obtain their water requirements at the right times of year, but a small amount is possible in those rain-fed areas which have well distributed precipitation. Double cropping is also confined to crops which can be harvested within five to seven months of the date of sowing, and apart from vegetables there are only a limited number of these.

In the 1950 agricultural census just under 100,000 acres were reported as carrying two crops and nearly one million acres in 1960. The 1970 census omitted information on this topic, but by using data published for the irrigation districts and making a guess for other areas, we may surmise that the area had risen to about two million acres in 1980.

As to the future, it is obvious that the expansion of double cropping cannot continue at the pace of the past thirty years. The combination of favorable ecology and ample, year-round water supplies is not found in many of the existing irrigation districts; still less is it present (and not yet exploited) in rain-fed areas. Therefore the chief expectation rests on how large a program of *new* irrigation can be accomplished in the coming years. If we suppose provisionally an additional 3.7 million acres of irrigation during this period (up to 1990) and if we recall that around 20 percent of the presently irrigated land carries double cropping, this should provide some 750,000 acres. But such a calculation would be unrealistic because most of the new irrigation will be located in the south where the local ecologies are much less favorable to double cropping practices. It would be more prudent to set a figure of 500,000 acres for these new areas, and to supplement this with say 250,000 acres representing possible further additions to double cropping in already established irrigation districts. Altogether we would thus reckon with a 750,000 acres gain for crop production from this source.

CONVERSION OF PASTURE

Another way of adding to the crop area would be by ploughing up some of the existing pastures. However, the country is already short of good grassland, and indeed many of the grazing areas are deteriorating because they are overstocked. Moreover, if it is desirable, as it is, to expand the livestock sector, it might not be prudent to encourage the ploughing-up of pasture unless new pasture areas were simultaneously being created elsewhere. As we have seen, this may indeed be a prospect in the southeast, and hence it may be legitimate to think of ploughing a limited amount of pasture in regions where the climatic and other conditions are sufficiently favorable.

It is not easy to form an opinion as to how much conversion of pasture to arable has actually occurred during recent decades. The census statistics show an increase in pasture area from 164 to 194 million acres between 1930 and 1970. The increases took place mainly in the northwest and in the far south, while declines were registered in the central states to accommodate increases in cropping. We may surmise that in those regions where increases in the arable area were large, the additions were obtained mainly from pasture and not so much from forest and other land. There may well have occurred an in-and-out movement, in the sense that forest and hitherto idle lands were being incorporated into pasture while simultaneously pasture land was being transformed into arable. But only detailed studies of particular regions would reveal the extent of this phenomenon.

In asking about the future possibilities for obtaining arable land from pasture, we have to remember that over two-thirds of the nation's grasslands are located in the arid and semiarid areas of the north, almost all of it on lands which are too marginal for cultivation without irrigation, though here and there a few small patches may enjoy circumstances sufficiently favorable for cropping. The chief potential must be looked for in areas of good rainfall, which really amounts to smallish areas along the southern Pacific coast and larger ones in the Gulf states.

To some considerable extent the rigidities in the land tenure legislation impede such conversions from taking place. In Tabasco we also need to maintain sufficient non-inundated grassland for the cattle population. In parts of Chiapas, Oaxaca, and Guerrero the changes ought to be in the other direction, turning back to pasture the marginal arable land which ought not to be being cropped. Only in the peninsular states have we identified a prospect of jungle clearance, a substantial part of which should probably be dedicated to grass. On the whole, therefore, the potentialities for expanding the arable area at the expense of pasture appear modest. No figures are suggested because the scope for such transformation has already been taken implicitly into account earlier in this chapter when we estimated the number of acres that could be won through programs of land clearance.

A SUMMARY OF POSSIBILITIES

The present chapter has been devoted to an exploration of the potentialities for expanding the area of crop production, leaving to later examination the prospects for raising the per-acre productivity through greater irrigation and other technical improvements. This exploration was conducted against the background of what has actually happened during the past fifty years, and taking into account the greater physical difficulties of continuing in the future what has successfully been accomplished in the past.

Between 1930 and 1960 the arable area, as noted earlier, increased by over 17 million acres, or 53 percent, constituting a growth rate of 1.4 percent per year. Of this additional area some 12.7 million acres were rain-fed land and only 4.5 million irrigated, most of the latter being obtained after 1940. It was also noted how little expansion was possible in the dry Altiplano during the latter part of the period, because virtually all the cultivable land was in use, including some submarginal land which should not be under the plough. Since 1960, and more probably at some date around 1965, between the fourth and the fifth agricultural censuses, the arable expansion came to an end and the area either stabilized or slightly declined (depending on how the statistics are adjusted and interpreted). But the nation's demand for food has not stagnated; it is growing at a faster pace than ever before in Mexico's history, which is why the food production challenge has acquired such a dramatic dimension.

There are various directions in which an increased output of food can be sought, and all of them will have to be exploited simultaneously and energetically, including as one of the more important the mobilizing of new land for cropping. In the light of recent experience this will involve the formidable task of arresting and reversing the recent trend at a time when all the more accessible and ecologically suitable land has already been brought into use. It is, therefore, a task which agronomically speaking needs to be undertaken with great skill and prudence, inasmuch as doing too much too rashly might be just as harmful as doing too little, because the former could conceivably cause irreparable damage to the country's last remaining untapped agricultural resource. It is a task which needs to be preceded by a far-reaching program of research to clarify the at present largely unknown techniques of soil management and fertility maintenance under tropical conditions.

Nevertheless, it would be incorrect to regard this as a purely technical problem amenable to technocratic and administrative solutions. It is much more than just a matter of bulldozers and large-scale organization. One has to ask why it is that the rhythm of expansion so suddenly came to a halt, because the answer to that question may to a considerable degree indicate the directions in which solutions could be sought. After all, farming is run by human beings, in Mexico several million of them, and human beings work in response to stimuli of various kinds. It may be that since 1965 there has occurred a weakening in the force of the stimuli which up till then had induced the farmers to continue expanding their cropland; or perhaps, while the stimuli remained unchanged, the task of expansion became more difficult both physically and financially; or perhaps both factors were operating.

One of the more important stimuli is the relationship between prices and costs—between the prices of the commodities which farmers sell and, on the other hand, the costs of the inputs which they use in production and the consumption goods which they need to maintain and if possible improve the living standards of their families. It may well be that this prices/costs relationship began to deteriorate just at a time when it was becoming more expensive per acre to bring additional land into cultivation (more expensive because the land was more remote, or needed more expenditure for bush-clearing, land leveling, irrigating, or draining). Farmers, even mini-farmers, are not as stupid as is commonly supposed. They are adept at making calculations of what is profitable for them, and they are quite capable of figuring out whether or not it would pay to bring into use an additional piece of land at the prevailing level of prices. To a large extent cropland expansion terminated because at existing prices, several of them ordained by government, it had become unprofitable.

Besides the regime of prices there has also been another group of influences which have played an outstandingly important role, namely those deriving from the land tenure system and the carrying out of the agrarian reform. Without any doubt, the agrarian reform in its early years provided a strong stimulus to agricultural expansion, in particular by putting large areas of mostly good quality but previously uncultivated land into the hands of a new class of operators who were eager to make it productive.

Yet a program which was highly beneficial under a certain set of circumstances may, if persisted in when the circumstances have changed, cease to have beneficial effects and may even begin to cause harm. For instance a large proportion of the new ejidos created during the past fifteen years have been given land which was being well farmed by its existing operators, but whose productivity declined because of lack of know-how among the new settlers. On the other hand, many of the owner-operator farmers have been desisting from making investments in improvements either because to do so would bring them in conflict with the agrarian laws or because they stood in continual fear of possible expropriation. In these ways, and in others which will be mentioned later, the land tenure system began to operate as a disincentive just when the nation was coming to need rapid growth of output, entailing stronger rather than weaker incentives to the farming community.

Faced with this combination of discouragements, it is not surprising that the farm people lost their dynamism, decided to choose in the main a passive role and wait to see what the government would do. In this situation different courses of action are possible. One, for instance, would be for the government to take the view that the tasks of expanding the food production base are too large and complex for individuals to undertake, and that therefore the major reponsibility must be shouldered by the public authorities. Another would be to reestablish a regime of effective stimuli to individual initiative by more generous adjustment to relative prices and by removing some of the less desirable rigidities in the land tenure system. Other alternatives could be envisaged consisting of various combinations of these two recipes. These matters will be considered later, but what has to be emphasized here is that without some bold decision making there is little prospect for operating successfully any sizeable programs for cultivating more land.

Technically the potentialities exist, even if they be not so grandiose as some people have alleged. We have identified some ten million acres in the southeast provided that agronomic research can tell us how to cultivate them without destroying their fertility. Of that acreage about one quarter could be in crops. There are probably another one and a quarter million acres scattered around the country. The combined total of 3.75 million would be equivalent to 2.2 million acres of harvested area. But without more systematic research and more flexible land tenure arrangements these goals will not be realized.

There are also possibilities, more modest but not negligible, for squeezing more acres of crops out of the existing arable area, and our tentative estimates are:

Reduction in fallow	750,000 acres
Reduction in losses	250,000 acres
More double cropping	750,000 acres

This total of 1,750,000 acres may be compared to an amount of some 4.2 million acres which was obtained from the same three sources between 1940 and 1970.

Altogether this amount plus the 2.2 million acres mentioned in the previous paragraph, gives us a total of almost 4 million acres or, since it is appropriate to be less precise in this type of projection exercise, let us conclude by saying that the potentialities for mobilizing more land may, when everything is taken into account, add up to something between 3.5 and 4.5 million acres which represents 9 to 11 percent of the presently harvested area. It would probably take 10 to 15 years to accomplish such a program. Although this estimate may appear disappointingly low to some readers, various considerations suggest that it will prove difficult to achieve so much in the time envisaged, because the physical obstacles are more formidable than has hitherto been supposed, the financial costs will be substantial, while the problem of mobilizing the enterprise and willing cooperation of the farming community who ultimately will make or break the projects, still remains to be resolved.

4. Irrigation

The story of the development of irrigation in Mexico provides a noteworthy and dramatic example of successful governmental intervention and investment in the farm sector. It is a story, in the main, of brilliant concepts brilliantly executed, which have brought great material benefit to large numbers of producers, both ejidatarios and private farmers, and which made possible the rapid expansion of production needed to match the demographic explosion.

In few countries has so much irrigation been achieved in such a relatively short period of time, and in few has the contribution to national food requirements been so decisive. By the end of the seventies Mexico had just over 20 percent of her crop land under irrigation, and this irrigated area was providing about 45 percent of the total crop production. In a country where a large proportion of the total area is arid or semiarid, where a large area is mountainous with slopes too steep for cultivation, and where a considerable area is subject to excessive rainfall causing erosion and flooding, the amount of land which can be cultivated successfully without artificial aids is relatively limited, and that is why irrigation has played, and will continue to play, such a vital role in Mexico's agriculture.

It has been estimated that in 1910, at the beginning of the Revolution, there were approximately 2.5 million acres of land under irrigation, of which some 1.7 million were cultivated.[1] Much of this had been created by various land companies, mainly American, with the intention of growing such crops as sugarcane and cotton. The Constitution of 1917 nationalized the country's water resources, and in 1921 an Irrigation Directorate was set up in the Department of Agriculture and Development, as it was then called, to be followed in 1926 by a National Irrigation Commission which was given responsibility for all irrigation affairs.

Among the early projects some were located in the central states, e.g., Aguascalientes, Guanajuato, and Hidalgo, a few in the north (Chihuahua and Nuevo León), while the areas in Sonora and Sinaloa were taken over from the

---

[1]Centro de Investigaciones Agrarias, *Estructura agraria y desarrollo agrícola en México,* vol. 3, México, 1970, p. 5.

land companies and amplified. However, in the 1930s progress was slow, an increase of only 375,000 acres according to the 1940 farm census.

But as the result of vigorous initiatives by the commission the pace of development quickened in the early forties. One of the first acts of President Alemán was to create a separate Department of Water Resources (1947) to which were transferred the responsibilities of the commission. Funds were allocated generously, so that by 1952 the annual rate of irrigation investment had nearly trebled, compared with 1946, and had nearly doubled in real terms.[2] This dynamism was maintained until the late fifties, and the 1960 farm census reported 8.7 million acres of irrigated land compared with 4.5 million twenty years earlier.

There are differences of opinion among historians as to what motivated this massive public investment in irrigation works. Probably a combination of influences was operating, each reinforcing the other, since no one alone would have been a sufficiently strong stimulus. First, the domestic market demand for food began its sharp upward trend around 1940, due partly to more rapid population growth and partly to urbanization. Secondly, the out-break of war in Europe in 1939, followed two years later by the entry of the United States, created a strong international demand and export opportunities for agricultural commodities at attractive prices.

Thirdly, the expansion of the crop area in the thirties had occurred almost entirely on rain-fed lands and had used up the best of these, so that further expansion in the forties and thereafter would need the supplementary assistance of irrigation. Fourthly, because Mexico's industrialization was beginning to acquire momentum, this created a need for imports of raw materials and machinery on an increasing scale, which could not be paid for solely by the traditional export of minerals but had to be financed to a substantial extent by other exports, among which agricultural products offered the best prospects. All these factors added up to a situation in which a policy of irrigation investment could be demonstrated to be overwhelmingly in the national interest.

At the end of the fifties these influences had worked themselves out, and the pace of development slackened. The statistical data of the next decade are somewhat conflicting. Probably the 1960 farm census overreported while the 1970 one underreported the irrigated acreage. A reasonable guess would put the real area in 1960 at around 7.7 million acres and that of 1970 at about 9.5 million acres. Between 1970 and 1980 the Department of Water Resources completed some 1.8 million acres of large and small works, though it is not known what proportion at the latter date was already operational.* Altogether, irrigation has become progressively more important, passing from less than 13 percent of the total crop area in 1940 to more than 20 percent in 1980.

---

*Early in 1977 the irrigation functions of this department were transferred to the Department of Agriculture.

[2]Navarrete, Ifigenia M., ed., *Bienestar campesino y desarrollo económico,* Fondo de Cultura Económica, México, 1971.

The geographical location of Mexico's irrigation projects has been dictated chiefly by the availability of water in or near arid or semiarid zones, having substantial areas of relatively flat land. The topography and climate of Sonora and Sinaloa provided classic sites for the irrigation engineer to exercise his talents: considerable rivers breaking through narrow mountain gorges (suitable for dams) and issuing out into wide coastal plains, where the natural rainfall was rather low so that the water supply could be almost entirely man-made and man-controlled. The irrigation of these two states, together with the Río Colorado district in Baja California, makes up nearly 40 percent of the total irrigated area in the Republic.

Major projects are also located in other northern states; notably the Lagunera near Torreón on the borders of Coahuila and Durango, the Ciudad Delicias works in Chihuahua, as well as two districts in the northeast on the Río Bravo and the Río San Juan respectively. In the center of the country the most important areas are those of Tula in Hidalgo, the Alto Río Lerma in Guanajuato, the districts in northern Michoacán also using waters of the Lerma basin, and lastly the Tepalcatepec project around Apatzingán. In the south and southeast few projects have been built, since the rainfall in most places is more than adequate.

At the last full count (1970) the irrigated area was divided almost equally between private farmers and ejidatarios, while during the seventies the latter probably slightly increased their share (Table 4.1). In the thirties and again in the sixties the increases in ejido irrigation were obtained largely by expropriating private irrigated land and transferring it to ejidos. In two senses, however, the private farmers are better placed: first, there are more irrigated acres per private farmer than per ejidatario, the latter group being so much more numerous; and secondly, of the private arable land almost one quarter is irrigated, whereas of the ejidatarios' only 15 percent is irrigated.

In the northwestern and northeastern regions private farm irrigation dominates over ejido irrigation, both absolutely in numbers of acres and in percent as a proportion of the total arable belonging to each tenure category. But in the center and southern regions the reverse is the case, reflecting the general predominance of ejidal tenure in those parts of the country.

The administrative organization of the irrigation districts was devised to take care of large areas of land cultivated by large numbers of occupiers, the typical situation in the northwestern and northern regions. For example, in Sonora the five main districts range between 45,000 and 540,000 acres each, while Sinaloa has four districts which average 315,000 acres each. Another ten districts extend beyond 100,000 acres each. However, at the other extreme many districts have less than 2,000 acres apiece and some are very small indeed; thus there are twelve districts each below 250 acres in size. Most of these small districts are located in the states of Jalisco and México. They came into existence for a variety of historical reasons, but each possesses the same elaborate administrative setup as the large districts.

On the other hand, the government also administers a large number of so-called "irrigation units," ranging in size from a few hundred to a few thousand acres. These are believed to cover almost one million acres and are

TABLE 4.1

IRRIGATION AREAS BY TYPE OF TENURE

(in thousand acres)

	1930	1950	1970	Irrigation as %of arable in 1970
Northwest:				
Private	712.3	948.2	1,863.1	67.7
Ejidal	21.6	714.4	1,411.6	37.7
North:				
Private	1,040.4	724.3	602.2	25.5
Ejidal	133.3	584.3	615.9	20.6
Northeast:				
Private	228.7	453.7	691.1	39.9
Ejidal	34.2	259.7	386.6	31.0
North-center:				
Private	91.1	94.7	169.1	10.6
Ejidal	31.8	94.3	138.4	5.5
West-center:				
Private	955.5	503.6	698.6	16.1
Ejidal	133.4	928.3	1,120.1	15.4
Center:				
Private	382.4	292.6	338.9	13.5
Ejidal	175.0	364.4	456.9	13.0
Gulf-south:				
Private	162.6	143.9	129.9	4.4
Ejidal	11.0	67.9	209.4	3.3
Peninsula:				
Private	30.5	11.1	11.3	3.6
Ejidal	0.0	1.1	10.5	0.8
Mexico:				
Private	3,603.4	3,172.3	4,504.3	24.4
Ejidal	540.7	3,014.4	4,349.5	15.1

SOURCE: Statistical Office, decennial farm censuses. Communities are included with private farms in 1930 and 1950, but with ejidos in 1970. The irrigation areas of communities are negligible.

located mainly in the central states. Unfortunately, these "units," unlike the districts, do not furnish annual reports to the department, so that little is known about their crops and output.

Still less is known about the remainder of the irrigation land, most of it in private hands. As to its extent the Water Resources Department and the census each publish different figures; probably the true figure is near 3.4 million acres, widely dispersed around the Republic. This area seems to have remained remarkably constant during the past fifty years, chiefly because private farmers have been unable to obtain permits to build dams or sink wells.

During the seventies the government placed considerable emphasis on building very small irrigation works, equipped either with earth dams or small capacity wells, bringing water to areas ranging from ten to a few hundred acres. These projects are designed and executed by small teams, each composed of technicians in the various disciplines required. As a result, important benefits accrue to poor farmers in zones of marginal land; but the totality of these projects does not, and is not intended to, make an important contribution to the national irrigation picture.

IRRIGATION INVESTMENT

The large and imaginative irrigation programs which began to bear fruit in the 1940s required what at that time seemed a great deal of money, though in retrospect the programs can be considered to have been remarkably cheap. It is indeed often the experience of nations that the taking of a bold expenditure decision is castigated as reckless by contemporary critics, but later when the investment is seen to have been indispensable it is also seen that its costs would have been many times greater if the decision had been delayed. This has certainly been the Mexican experience, and maybe the time is ripe for making new investment decisions in agriculture, the justification for which will only be fully perceived as the future unfolds.

In the days of President Camacho (1940–46) irrigation works accounted for 97 percent of all public investment in the farm sector, and the proportion has barely fallen below 80 percent in any subsequent presidential term. Measured at constant prices, the volume of irrigation investment has increased in every six-year term, with particularly large increases in the periods of President Alemán (1946–52) and President Díaz Ordaz (1964–70).

A disturbing feature has been the steep rise in investment outlays per acre served. In real terms this cost increased more than sixfold between the forties and the seventies, which confirms our previous observation that, had the carrying out of the programs been delayed thirty years, the burden on scarce national investment resources would have been much more severe.

Two principal factors have accounted for this dramatic increase in capital costs. One has been the steady increase in the relative prices of project inputs, not so much with respect to earth-moving equipment and other machinery as in the relative price of manual labor (a phenomenon welcomed as a sign of advancing prosperity); and because irrigation works are labor-intensive, this has been an influential factor.

The second principal factor has been the exhaustion of all the more propitious sites for dam building and canal construction, with the result that new projects involve more costly works for creating reservoirs, longer canals for bringing the water to the farm land, heavier expenditure for canal lining as well as for drainage since most of the new projects happen to be located in zones of abundant summer rainfall. All these factors will continue to influence adversely the cost of irrigation investment in the coming years.

Fortunately, since the early seventies the World Bank and the Interamerican Development Bank have taken a strong interest in agricultural investment, and especially in irrigation investment. Mexico has benefitted substantially from this orientation of their policies; in fact most of the Mexican government's new irrigation investment has been financed by loans from these two bodies (for instance a large World Bank loan for rehabilitation works in the Bajo Río Bravo district). These two institutions will probably continue to participate in programs of this nature, provided the financing remains within the framework of the government's foreign borrowing policies.

It would be interesting if our brief review could be supplemented by some data on private investment directed toward the same objectives. Unfortunately nothing is known on the subject beyond the fact that the area of privately operated works has not changed for many years, mainly because new water exploitation permits are not granted to private individuals. Certain older installations on private irrigation may have been replaced, but otherwise private investment is confined to repair and maintenance of existing facilities.

PRODUCTION FROM IRRIGATED AREAS

In the early stages of Mexican irrigation, cotton, corn, and wheat were the chief crops, and together they accounted for 85 percent of the harvested (irrigated) area.* Up to the mid-fifties cotton continued to occupy over half the irrigated land, but gradually the newer crops—safflower, soybeans, and grain sorghums—began to make an appearance and, together with sugarcane, took over much of the land formerly used for cotton (Table 4.2). During the 1970s corn, wheat, oilseeds (as a group), feed (alfalfa and sorghums), and cotton each accounted for about 15 percent of the area, while minor crops were becoming increasingly significant. All this constituted a process of diversification toward a wider range of crops in response to a more diversified market demand.

Since in the climate of most of Mexico, annual crops thrive better if given additional water, there is considerable competition between crops for the limited amount of water available. The determination of the pattern of land use depends on several factors. It results partly from changes in the relative profitability of different crops, which in turn is affected by technical progress, for instance in plant genetics, in the devising of appropriate fertilizer mixtures, and in pest and disease control. It results also from the ecological characteristics of different regions: the adequacy or otherwise of the water supply to the reservoirs, the prevalence of specific pests and diseases which prove hard to eradicate (as in the case of cotton cultivation which ultimately had to be abandoned in the northeast and in Michoacán).

---

*For lack of information about irrigated land outside the so-called "districts," the discussion in the following pages will refer to these districts, unless otherwise specified. Data on the other areas could theoretically be derived from the difference between the national figures for irrigation and the district figures. Such a calculation indeed shows that the additional lands are devoted mainly to corn, beans, sugarcane, and alfalfa, which appears plausible. However, other discrepancies between the two sources are so great that it would be imprudent to utilize the results obtained by this indirect approach.

TABLE 4.2

**DISTRIBUTION OF HARVESTED AREA OF SELECTED CROPS
IN IRRIGATION DISTRICTS: 1944–45 TO 1978–79**

(in percentage of total area harvested)

Crop	1944–45	1954–55	1964–65	1974–75	1977–78 and 1978–79 average
Corn	24.9	10.9	23.8	13.7	15.7
Beans	0.4	1.6	0.2	5.5	3.1
Wheat	19.3	22.3	25.5	16.3	14.0
Rice	0.6	2.1	2.6	4.0	2.1
Sesameseed	0.6	0.4	1.6	1.1	2.1
Peanuts	0.7	0.3	0.3	0.2	0.2
Safflower	0.0	0.0	0.7	9.1	9.8
Soybeans	0.0	0.0	0.8	7.8	6.8
Alfalfa	3.5	1.8	3.0	1.7	3.4
Sorghum	0.0	0.0	5.3	15.2	13.0
Peas	2.7	1.0	1.4	1.3	2.9
Tomatoes	0.8	1.2	0.9	1.0	1.1
Cotton	41.5	53.1	21.8	6.0	8.9
Sugarcane	a	a	3.9	3.5	3.6
Other crops	5.0	5.3	5.2	13.6	13.3
Total	100.0	100.0	100.0	100.0	100.0

SOURCES: "Agricultural Statistics of Irrigation Districts," published up to 1977 by Department of Water Resources and thereafter by Department of Agriculture, Bureau of Agricultural Economics.

aIncluded with "Other crops."

Not all irrigation districts shared in this crop diversification. The least occurred in the northeast where cotton had to be replaced by two low-value crops, corn and sorghums, to the virtual exclusion of all else because of insufficient water available within the irrigation district. On the Río Colorado cotton declined to a third of the total area and farmers turned to alfalfa. In Sonora and Sinaloa wheat, safflower, and soybeans became profitable, and most expecially tomatoes for export.

Crops grown under irrigation have higher per-acre yields than rain-fed crops, not only because of the water but also because Mexican farmers in the irrigation districts tend to practice more modern technology than those elsewhere. For one crop the yield difference may be twofold, for another as much as threefold. Thus more than 80 percent of the wheat, cotton, soybeans, and safflower are irrigated and some two-thirds of the rice, chiles, and tomatoes. In general the yields under irrigation compare favorably with yields under similar conditions in the United States. Some, however, are notably worse in Mexico: corn, tomatoes, and grapes, for example. In a few crops

TABLE 4.3

**Per-Acre Value of Output of Selected
Crops in Irrigation Districts**

(in dollars per acre)

	1960–61 and 1961–62 average	1966–67 and 1967–68 average	1972–73, 1973–74, and 1974–75 average	1977–78 and 1978–79 average
Tomatoes	176.28	679.78	1,348.38	2,128.94
Cotton	139.51	206.15	427.81	547.06
Alfalfa	174.48	276.34	354.84	531.72
Peanuts	63.14	102.39	266.37	394.48
Peas	42.84	52.61	160.10	365.08
Sugarcane	139.38	183.30	261.34	336.10
Rice	90.16	124.18	337.58	282.81
Wheat	75.13	84.22	180.98	211.28
Soybeans	75.39	97.84	227.58	208.30
Beans	63.30	87.42	177.91	169.71
Safflower	56.83	69.18	169.32	165.79
Sesameseed	33.79	64.02	145.59	157.18
Barley	70.56	98.60	159.84	152.61
Corn	43.53	63.89	117.26	141.80
Sorghum	42.06	59.02	141.70	136.37

Sources: See sources for Table 4.2.

Note: The price index of farm inputs (see Chapter 9) moved as follows: 1960 =100, 1967 =109, 1974 =154, 1978 =296.

Mexican yields far surpass those of the U.S.A., e.g., wheat and cotton; but then not much U.S. wheat or cotton is irrigated.

Yields can also be measured in value terms of course, and it is useful to note the large contrasts which prevail under the conditions of the Mexican market (Table 4.3). In 1979 when tomatoes yielded some 2,000 dollars per acre and grapes around 1,000 dollars, corn produced less than 150 dollars and wheat and beans little more. The reader should remember that these figures have nothing to do with profitability, since they indicate gross receipts and tell us nothing about differences between crops in production costs. Obviously, tomatoes and grapes, being labor-intensive, involve much higher per-acre costs than do wheat and corn.

Dollar output per acre is also a useful yardstick for measuring the relative performance of irrigation districts in different parts of the country. Table 4.4 presents the gross value of output, averaged over three years, in the eleven largest irrigation districts, and shows a fourfold difference between the most and the least productive. The results depend largely on which are the predominant crops in a locality. Thus, Culiacán does well mainly because of

TABLE 4.4

**AVERAGE PER-ACRE VALUE OF TOTAL CROP OUTPUT
IN SELECTED IRRIGATION DISTRICTS**

District	1976–77 and 1977–78 average (dollars per acre harvested)	District	Increase in real per-acre value between 1951–52 & 1977–78 (1951–52 = 100)
Culiacán	508.36	Alto Río Lerma	263
Lagunera	406.56	Bajo Río Bravo	210
Valle del Fuerte	297.88	Valle del Fuerte	159
Río Colorado	296.74	Río Mayo	154
Ciudad Delicias	271.84	Río Colorado	149
Alto Río Lerma	261.69	Lagunera	148
Río Yaqui	244.06	Río Yaqui	144
Río Mayo	230.15	Culiacán	143
Bajo Río San Juan	139.02	Ciudad Delicias	138
Bajo Río Bravo	92.21	Bajo Río San Juan	80
Average of all irrigation districts	256.74	Average of all irrigation districts	150

SOURCES: See sources for Table 4.2.

NOTE: The "real" per-acre value includes a correction for an inflation of 618.6 over
the 1951–52 price level (= 100).

its tomatoes, the Lagunera because of cotton. The Río Bravo and Río San
Juan come last because of their concentration on sorghum and corn.

From the analysis of individual crops we turn to consider the contribution which the irrigation area as a whole makes to the nation's agricultural
output. For this purpose, since nothing definite is known about the productivity of the areas outside the districts, we have arbitrarily assumed that their
gross output per acre averages the same as in the districts. We also assume
that the area of "ex-district" irrigation has remained constant at around 3.4
million acres.

Using these assumptions, it appears that the crop output from irrigated
areas, reckoned as a percentage of total crop output, rose from about 40
percent in the early fifties to 50 percent in the early sixties, since when it has
fluctuated above and below 45 percent. The two determining components of
this contribution are of course area and yield.

The share of irrigation in the total harvested area rose from less than 20
percent to some 25 percent in the sixties and even a little higher in the
seventies, because of the stagnation in the rain-fed acreage. Whether this new
trend will persist is difficult to determine.

As to yield, the statistics indicate that on average the irrigated land
produces about two and a half times as much as rain-fed land, measured in
value terms, the annual variations being from twofold to threefold. It is

generally supposed that irrigation has the advantage of reducing climatic hazards, compared with crops dependent on erratic rainfall. The data from 1961 onwards do little to support this supposition, showing only a slightly smaller deviation from output trend in irrigated areas than in rain-fed areas. This is partly because some rain-fed cultivation is located in areas of reasonably reliable rainfall and partly because many irrigation districts suffer either from water shortages in certain years or are periodically subject to cyclones or floods.

Over the years the productivity of irrigated land seems to have increased less rapidly than that of the rain-fed. This is quite normal: irrigation productivity started from a much higher level of technical sophistication and physical yield; the rain-fed was predominantly subsistence farming where there have been opportunities to change to commercial farming, at least in some regions. This may also suggest that the fruits of research may not have benefitted the irrigated areas so exclusively as is often alleged.

This examination of the role of irrigation in Mexico's farming may be concluded with one final observation. Although the figures show its contribution to total crop output to vary between 45 and 50 percent, its contribution to *marketed* output is much higher (the part which goes to supply the cities). A significant amount of rain-fed production is retained for the needs of the subsistence farmers' own families, whereas only a minute fraction of irrigation production is consumed on farms. Allowing for this, probably the irrigated areas account for over two-thirds of the marketed supplies of crop products; and therefore it is of strategic importance that these areas continue to function well, that where necessary they be improved, and that new irrigation works be programmed.

LAND TENURE AND PRODUCTIVITY

The questions that must now be posed are: who are the operators of the irrigated land and how much have they benefitted? When one bears in mind that these operators fall into two broad categories, the ejidatarios and the private farmers, the questions have considerable political significance, inasmuch as the debate concerning the two forms of tenure is still far from terminated. However, in examining the relative productivity performance of these two groups, we are again obliged to focus on the irrigation districts for the lack of more comprehensive information.

In 1975 there were just over 400,000 farm operators in the irrigation districts, a number which had doubled since 1955. Of these operators about 70 percent were ejidatarios, a proportion which has remained constant for many years. It also happens that ejidatarios constitute about 70 percent of farm operators on all types of land, so that their numerical representation in the districts may be considered normal.

While the number of operators doubled, the land area in the irrigation districts increased somewhat less, so that the size of the average production unit diminished. The average holdings of private farmers declined from

37.2 to 31.5 acres; meantime the average among the ejidatarios rose from 11.2 to 12.2 acres. These may not be very large changes for a time span of twenty years, but they are changes in the direction of more egalitarian land distribution.

At the time of the creation of any particular district, the policy in most cases was to set aside sufficient land to enable each ejidatario applicant to obtain between 12.5 and 25 acres, whereas the private farmers of the locality had divided among them whatever remained, subject to not exceeding the legal maximum of 250 acres per person. Since there were usually more applicants for ejidal land than could be accommodated, a certain degree of overcrowding occurred from the early days but did not occur so much among the private farmers. Nevertheless, in relation to the controversy about compulsorily reducing the maximum permitted size of private holding, it may be noted that already in the irrigation districts 85 percent of the private farms are below 50 acres.

As to changes through time in farm size, among the ejidatarios the proportion represented by units of less than 12.5 acres has considerably declined, a quite opposite trend from that in the rain-fed areas, as we shall see later. Among the private farmers the opposite has occurred, namely a reduction in the proportion of farms exceeding 25 acres. It seems unlikely that either of these changes has been caused by rearrangement of farm boundaries in the already existing districts. What seems more probable is that in the more recently created districts the ejidatarios have been allotted larger units and the private farmers smaller ones.

When we compare the several main irrigation districts, significant differences in farm size emerge. Generally speaking, the farm sizes are larger in the north and northwest than in the center, and naturally they tend to be larger in districts where private farmers outnumber ejidatarios. In the northwestern districts the private irrigation farms range from 50 to 150 acres and the ejidatarios' from 20 to 40 acres. In the center region the private holdings average 30 to 40 acres and the ejidatarios eight to 15. With two exceptions the average 30 to 40 acres and the ejidatarios' 8 to 15. With two exceptions the national average, which indicates that the remaining 127 districts excluded from this commentary, most of which are very small, also tend to have very small farms.

From the viewpoint of the individual operator, income depends not solely on his acres but also on their productivity. The overall per-acre productivity of the different districts has already been discussed, but now the matter must be looked at from the tenure angle. In the Mexican literature it is commonly stated that the ejidatarios are more ignorant, dispose of more meager financial resources and achieve lower levels of output per acre than the private farmers. The evidence only weakly supports this view. Taking an average of all the districts, the private farmers' output per acre in value terms was only some 10 percent above that of the ejidatarios in the late seventies. However, in the northwest, center, and southern regions, the differences were larger, chiefly because the private farmers in those regions cultivated higher value crops.

In the mid-sixties, when this type of information was first recorded,[3] the ejidatarios performed fractionally better than the private farmers. This implies that the private farmers have progressed somewhat more rapidly than the ejidatarios, but the contrast is not great. The relative performance of the two groups varied from district to district through time. Thus the ejidatarios registered more rapid progress in the north and northwest; elsewhere the situation was reversed. Generally, in cases where for one or another reason the cultivation of a high value crop had to be reduced or abandoned, the private farmers showed greater adaptability in turning to other high value crops.

On this topic, however, one important reservation has to be made. In many irrigation districts it is customary for ejidatarios to rent their land to private farmers. In those of the northwest, for instance, it is thought that 50 to 70 percent of their land may be rented out. Since these arrangements are quite illegal, they are not recorded; in the statistics the rented-out land and its output is attributed to the ejidatarios. If it be supposed that the technical efficiency of the private farmers is on average superior, achieving higher per-acre output, then part of this shows up as ejidatarios' output and unduly inflates their performance. The influence of this distorting factor cannot be measured but may be considerable. Subject to this one qualification, it would seem to be a valid conclusion that in zones where ejidatarios have a reasonable amount of land and employ sufficient water and other necessary inputs, they are just as capable of securing high land productivity as any other type of farmer. Where they fail is on their millions of acres of marginal land in arid and semiarid areas, where anyone would fail.

PHYSICAL PROBLEMS

Not only in Mexico but in other countries, as also in past civilizations, irrigation systems tend to pass through cycles of growth and decay: an initiation phase when the cultivators are learning what crops to grow and what techniques to apply, then a phase of full maturity when the area attains a high level of output, and subsequently a period of decline brought about by physical and man-made factors. In the modern world the cycle may be completed in a few decades whereas formerly it took centuries.

In Mexican irrigation a number of disquieting features can be observed. One of the troubles is the serious loss of water on its way from the reservoir, or river offtake, to the fields. When the irrigation engineers measure these losses they find that in the efficient systems some 60 percent of the water leaving the reservoir reaches the crops, and unfortunately few systems are this good. In the less efficient the figure is around 25 percent.

In several districts, though the soils are highly permeable, the canals have not been lined, causing losses through seepage. Some main distribution canals are up to 50 miles in length which results not only in seepage but in evaporation losses as well. In some cases the canal length is unavoidable due

---

[3] Water Resources Department, Statistical Report, No. 32.

to physical features of the terrain, but in other instances political decisions caused the works to be planned on a scale too extensive in relation to the available quantity of water.

Another cause of wastage is the inadequacy of land levelling on the cultivators' plots, so that one portion may be high and dry while another is excessively inundated. There is also the problem of educating the farmers in applying controlled amounts of water instead of practicing flooding as they were accustomed to do in the days when they used simple derivations from their rivers. In one Sinaloan district, now highly efficient, it took the supervising engineer ten years to cure the cultivators of those wasteful habits.

A further problem is the gradual silting up of the distribution system. Where a reservoir has been built in an erosion zone without an accompanying erosion control program (tree-planting, etc.), its cubic capacity can be materially reduced in very few years. In the distribution canals solid matter is deposited either from the water itself or from crumbling of the banks. In the absence of regular maintenance, which is often lacking, the system becomes partially choked, and whole sections of a district go out of cultivation.

A different problem which Mexico faces is the progressive salination of the land, caused either by infiltration from the sea or by faulty drainage. In some northwestern districts infiltration has become serious where the works were pushed too close to the coast (Sinaloa), or where the rate of water extraction has exceeded the rate of recharge of the aquifer (Costa de Hermosillo), and the water table has fallen. As to drainage, some systems never provided any, assuming that natural porosity would be sufficient, or the drains had insufficient capacity, or they became blocked through neglect of maintenance. It is hard to find any cases of sanctions having been applied to cultivators who have neglected their responsibilities in this respect.

Yet another problem afflicting several districts is the almost perennial insufficiency of the water supply. In some instances this resulted from miscalculations in the original design (where the engineers lacked adequate hydrological data). In others the local political pressures combined with an excessive number of applicants resulted in decisions to give a little water to a lot of people rather than adequate irrigation to fewer people.

Oddly enough, districts can be found where the problem is the reverse, namely frequent and unwanted flooding of the crop land which either prevents sowings or occasions losses before harvest. Flooding may be caused partly by faulty drains, partly by seepage in unlined canals, but the principal cause is heavy tropical rainfall concentrated in periods of just a few weeks or even a few days. In the northeast the flooding is such that field work is possible on only half the days of the year, and for only about 36 days during the critical months from August to November.

It is difficult to quantify the recurring damage resulting from these various physical deficiencies. Certainly the most extreme case is where land goes entirely out of cultivation. Strictly speaking, irrigated land should be carrying at least one crop per year, and in many cases two or more; yet in the Mexican irrigation districts it has been calculated by the Water Resources

Department that in a normal year about one million acres carry no crop at all. To this damage must be added the shortfall in per-acre yields resulting from one of other of the difficulties mentioned; and in some locations the cultivation of certain water-hungry though valuable crops has had to be prohibited. Of course, no irrigation system functions at its optimum level; nonetheless, there is reason for being disturbed by the deteriorating situation in many of the Mexican districts.

HUMAN AND ADMINISTRATIVE PROBLEMS

Strangely enough, the education of small farmers in correct water use went for many years entirely unsupported by financial inducements to economize water. Farmers paid for water according to their acreage, not according to the quantity of water used. Furthermore, the charges were so low that they covered only a minor part of the operating costs of the projects—in some districts only one tenth. It has seemed puzzling that successive governments pursued this policy of subsidizing irrigation water, to the benefit of what is after all the richest section of the farming community. However, during the presidency of López Portillo cautious steps were taken to modify this policy, payment quotas were increased, and in several districts water meters were installed so that users could be charged by volume.

Another problem is the incapacity or unwillingness of many farmers to carry out the tasks for which they are legally responsible, e.g., cleaning and repairing the secondary canals and drains, levelling their plots, etc. Although it would augment their crops and incomes, most small farmers refuse to undertake this type of work unless they are paid for it, just as they are accustomed to being paid for work on rural roads. Not infrequently this neglect also damages the irrigation facilities of neighboring farmers; yet no one is fined, and certainly no one is evicted.

The committees which direct the operations in each district contribute another problem with their inadequate functioning. These bodies are supposed to be representative of all the classes of cultivators in the district, together with officials from the government agencies concerned; but in practice the desirable participation is seldom secured because of local jealousies, political rivalries, or mere inertia. Mutual confidence is not improved when a committee, perhaps clumsily interpreting instructions from the government, seeks to impose the cultivation of certain crops which are technically unsuitable or economically unprofitable in that locality.

Awkward problems arise when water supplies fall below requirements, a topic upon which the articles of the Federal Water Law are ambiguously vague. Thus, one committee will distribute an equal quantity of water to each user, irrespective of the size of his farm, another will allocate according to acreage sown, while another will blatantly favor the local boss and his friends (this latter chiefly in the smaller and more remote districts). Also, in these latter situations the bureaucratic reporting procedures are unnecessarily complex, being identical with those devised for the 200,000 acre districts.

REFORM

The chief purpose of reforms on the human and administrative side should be to improve the efficiency of water utilization and the agricultural performance in the existing irrigation districts. As usual in tackling such problems, it will be necessary to devise an admixture of sticks and carrots. The most constructive single step would be to adopt throughout the system a policy of full payment for the cost of water supplied. This would entail equipping all farmers with water meters which they could well afford to pay for, and setting the price so as to cover normal operating expenses plus some amortization. Admittedly, a few districts are so "abnormal" and run down that such payments could not also cover the cost of rehabilitation. These the government will have to bear to compensate for past negligence. A useful consequence would be to give the cultivators a stronger inducement to maintain their canals and drains and level their plots.

Another reform, long overdue, would be to issue instructions to every committee to take steps to halt the fall in the water table of the district. Each well should have its meter, and continuing studies of the aquifer would show the safety limits of pumping. Exceeding these limits should invoke severe sanctions. Periodic consultations between engineers, agronomists, and farmers should determine what changes in cropping patterns would be appropriate in the light of prospective water supplies.

This leads naturally to the mention of alternative irrigation techniques: furrow irrigation using less water than general flooding, sprinkler irrigation offering still greater economies (although in conditions of low atmospheric humidity evaporation may be high), and drip irrigation, which is the most economical of all but is applicable only to high value crops because of its high installation costs. In addition, Mexico might give more attention to supplementary in place of full-scale irrigation, since in many zones the former would be sufficient for such crops as corn, sorghums, peas, and beans. By thus stretching available water over a larger area, production could be significantly increased.

On the administrative side, much remains to be done to reduce overlapping of the activities and programs of government agencies. In spite of the incorporation of Water Resources into the Agriculture Department, a number of parallel programs remain operative, often in competition with one another. Insofar as the duplication cannot be wholly eliminated, at least regular consultation between program directors at local level should be obligatory.

Finally there remains the problem of the physical rehabilitation of the more deteriorated districts. Since a number of these are forty years old or more, it is natural that major investment in rehabilitation has to be programmed. With some areas totally abandoned and with reduced yields in other areas, rehabilitation projects could probably increase the nation's food production by something like 15 percent.

A concrete example is District No. 25, the Bajo Río Bravo, where in the late seventies a major rehabilitation program was initiated. This had been one of the most neglected districts, with 75,000 out of its 500,000 acres out of production and a further 175,000 acres suffering from salination. The program, extending over seven years, envisaged repairing nearly 2,000 miles of canals as well as providing more ample and efficient drainage. It was expected to cost, at 1979 prices, some 350 million dollars, but apart from recovering abandoned land it would allow the farmers to diversify into more lucrative crops.

This example gives some idea of the magnitude of the expenditures which will be involved in rehabilitating the irrigated lands. The Bajo Río Bravo is not the only district in need of such a program, though it may be one of the worst. At a conservative estimate, something like 2.5 million acres will need investment in rehabilitation before 1990. And to delay the expenditure until later will only increase the ultimate cost, because deterioration is continuing each month and every year.

NEW LARGE-SCALE PROJECTS

Parallel with rehabilitation activities is a continuing government program for bringing irrigation to new areas of land. In order to set the various phases of this program within a long-term framework, the Department of Water Resources published in 1975 a "National Water Plan," which brought together on the one hand the future demands for water—human, industrial, and agricultural—and on the other the physical supplies and the projects needed to mobilize these. With regard to future programs for irrigation, the plan envisaged three broad categories: new major works, small rural water projects, and separately the rehabilitation tasks already discussed above. Its time horizon was to the year 2000. This plan still provides the main governmental framework.

As to the new major irrigation works, the plan distinguishes three groups: those which have reached the stage of design and execution, those which are the objects of feasibility or prefeasibility studies, and finally those at the stage of preliminary identification. In most cases the plan indicates the acreage to be benefitted and the probable investment cost.

In what follows we have eliminated projects whose main purpose is to improve drainage on existing farm land (especially areas in the south subject to annual flooding), and also those still at the identification stage, since they could not become operative before 1990. We are left with projects which would serve some 2.5 million acres, plus another half million acres of projects already in course of construction. (Between 1975 and 1980 a few of these latter were terminated.)

The geographic distribution of these three million acres is as follows: Northwest, 1.15 million acres; Northeast, 0.50; Center, 0.25; South & Southeast, 1.10. It will be seen that more than two-thirds will be concentrated in the northwest and the southeast, the remainder being widely distributed about the country. Such a distribution has both human and agronomic implications.

From the human angle, it is unfortunate that comparatively little new irrigation can be envisaged for the densely populated agricultural regions in the center of the country, but this physical limitation has long been recognized

and could not be overcome except at exorbitant cost. The major water potentialities are in the northwest and the southeast where the agricultural population is relatively sparse. In this context the government has stated that people
must be brought to where the water is and not vice versa.

The agronomic problems are of two kinds. First, although the northwest
possesses ideal climatic conditions for irrigation, the northeast and southeast
are regions subject to tropical rainfall and cyclones, giving rise to floods and
heavy crop losses. Secondly, as will be seen later, the future expansion of
Mexico's food demand will be oriented toward temperate zone crops much
more than toward tropical crops. For example, zones well adapted to coffee
and bananas will not grow wheat. This suggests caution in translating engineers' projections into additional agricultural production.

One further qualification is in order. The figures which water engineers
use for "areas benefitted" include land to be used for roads, housing, and
other non-agricultural purposes. Past experience suggests that out of the three
million acres programmed, only about 2.5 million could be available for
cropping. Furthermore, the plan envisages a level of construction activity
which would exceed by fifty percent the highest performance attained in any
previous decade. It is more than possible that shortages of skilled personnel
and of equipment might be experienced.

In any case a very real constraint may be the investment costs of the
plan. No overall budget was prepared, as is understandable in a program
scheduled to last 25 years, yet price tags were attached to individual projects.
Using data on construction costs in the late seventies, it may be surmised that
the investment in major irrigation projects would range between 2,500 and
2,800 dollars per acre at 1980 prices. In some of the awkward projects of the
southeast the figure might be higher. For a total area of 3 million acres, this
would imply an investment of 7.5 to 8.4 billion dollars. That is perhaps why
the plan has been referred to as "the engineers' dream and the politicians'
nightmare."

SMALL-SCALE IRRIGATION

The National Water Plan also makes provision for small-scale irrigation
works which in many instances also include drinking water for villages and
their drainage. While it is difficult to calculate with precision the likely
consequences of these, it would seem that during the eighties some 300,000 to
400,000 acres might obtain irrigation via these projects. Although the engineering works involved tend to be much more modest than in the major dam
projects, the administrative costs appear particularly high; indeed these latter,
not fully budgeted in any of the programs, are implicit in the bureaucratic
structure of the agencies responsible.

This is a cogent reason for asking whether the present procedures of
governmental irrigation programs are the only possible way of augmenting the
area, whether there may not be simpler ways of getting some of the tasks

accomplished, at least as a supplement to the official activities. With respect to digging a well, fitting a pump, even constructing a small earth dam, there are thousands of individuals willing, eager, and competent, if only they were permitted to undertake such works.

One example from the state of Chiapas will illustrate what is possible. A man bought a three-inch pump for 2,000 dollars and installed it on one of the many small rivers in a valley near Villa Flores. With the aid of two helpers he constructed a miniature canal lined with plastic shopping bags, and with total construction costs of 1,000 dollars, mostly wages, he irrigated 50 acres on which he cultivated gladioli and other flowers, recovering his outlay several times with one year's production. This is economic irrigation correctly located where the production potentialities are high and executed without any bureaucratic overheads.

It is believed that in that region of Chiapas there may be a quarter of a million acres to which similar treatment could be applied. In the state of Veracruz there are substantial areas having a like potential. Even in parts of the central plateau which lack rivers the groundwater supplies have never been fully explored since much of the water lies in limited concentrations too small to be of interest to large government agencies. Many ranchers could stock more heavily if they were permitted to grow a certain quantity of irrigated feed crops to carry their animals through the dry season.

These water resources and the capacities of these farm people are almost totally neglected, with substantial loss to the nation in terms of agricultural output. Why are they neglected? On the one hand the major government agencies are interested only in large works to which prestige can be attached. The small official programs are hamstrung for lack of finance and by bureaucratic rigidities. On the other hand, the individual farmers who would like to irrigate face insuperable obstacles: the virtual impossibility of obtaining a water utilization permit, the tenure problems of ejidatarios, the cattle rancher's fear that his property will be expropriated if he engages in crop production, etc. Individuals with a lifetime of local experience are frequently in a better position than water engineers from Mexico City to judge what pieces of land can be safely and profitably irrigated, and they will contrive to carry out the works at one tenth of the cost the government agencies incur.

It is difficult to put forward even an approximate estimate of the potentially irrigable area, assuming that these barriers could be removed. Many favorable sites are known to exist up and down both the Gulf and the Pacific coasts; probably a number of opportunities also exist even on the Altiplano. It is entirely likely that before 1990 one million acres could be brought into cultivation by these means. No government funds would be involved, there would be no large and costly failures (any miscalculations being paid for by the sponsoring individuals), the new projects would become operational much more rapidly than those subject to the time-taking, decision-making procedures of the bureaucracy. Of course, this approach has its limitations; it is unsuited to the really large projects, often multipurpose in character. But as a fruitful adjunct it should be explored to the full.

THE LONG-TERM OUTLOOK

The conclusions which we have reached are that within the decade of the eighties the major works that have been programmed could add some 2.5 million acres to the effectively irrigated crop area, while an additional 300,000 to 400,000 acres might be won through the small-scale official projects.* This would constitute a much faster rate of expansion than has ever been achieved previously in Mexico in a similar period.

Bold though these programs undoubtedly are, they do not appear so adequate if measured against the expected growth of population and of consumer demand. Thus in the year 1930 there were approximately four Mexicans for every acre of irrigation; what has happened since then and what may be expected in the future can be seen from the following figures:

Year	Mexicans per irrigated acre
1930	4.0
1975	6.0
1990	6.8
2010	7.0

Despite all the efforts and investments, population is growing more rapidly than the irrigated area and will continue to do so even if the foreshadowed programs are fully implemented.

Furthermore, the Water Resources Department estimated that the theoretical ultimate limit to irrigation in Mexico is about 23.5 million acres and that the practical limit is closer to 21 million acres. Assuming that this latter goal were pursued and achieved by the year 2010, the position would still have worsened to seven persons per acre. This is a sobering thought which will stimulate us to inquire in subsequent chapters what other developments, technical and otherwise, might be envisaged that might make the production outlook appear less gloomy.

However, the gloom does not end here. To obtain a complete picture of the production potential we need to consider the rain-fed areas as well, utilizing the projections made in the previous chapter. To do this it has been assumed that each acre of irrigation equals two-and-a-half acres of non-irrigated land, and the results have been calculated in terms of "acre units" of rain-fed land. When these are related to the population at different dates, the following picture emerges:

Year	"Acre Units" per capita
1930	2.4
1970	1.2
1990	0.7

The Chapter 3 projection was for four million additional acres by 1990. For those who consider this estimate too conservative, one may note that even an

---

*Although we have also just mentioned a further one million acres which could be irrigated through private initiatives, the conditions which would permit this are unlikely to be fulfilled.

expansion by 7 million acres would raise the 1990 man/land ratio to merely 0.75. Thus we are forced to the conclusion that the food production resource base of Mexico is diminishing and will continue to diminish, unless unexpected changes can be brought about in the rate of growth of population.

Faced with such a conclusion it would be foolish to carp at the size of the investment bill that may be presented in striving to achieve these objectives. This does not mean condoning reckless and extravagant expenditure, even though more ample funds may become available from the ever-expanding petroleum exports. It does mean that large irrigation investments will have to be budgeted even while food imports are inevitably increasing. It probably also suggests that the government might think again about that million acres of irrigation that could be had by removing the institutional obstacles to private initiative.

Howsoever this may be, the broad conclusion remains that the nation will have to run forward very fast in order not to slide backwards too rapidly in terms of per capita agricultural resources. Whatever is accomplished in land mobilization and in irrigation, the man/land ratio will worsen. The degree of this "worsening" will depend on the amount of human and economic resources which the nation decides to dedicate to these tasks, on the imagination with which the programs are conceived, and on the efficiency with which they are executed.

5. Livestock Expansion

The pre-Hispanic Mexicans had little experience in breeding animals for food, while the interest of the Spanish invaders was limited to horses. As a result, for three hundred years farming in Mexico continued to be dominated by crop growing. But when in the second half of the nineteenth century the Mexican government made large grants of land to U.S. citizens and others, especially in the northern regions, these new owners proceeded to develop cattle ranching along the lines to which they were already accustomed from Texas to the Dakotas. One consequence was that the periodic upsurges of anti-Americanism came, by association of ideas, to include ranching.

In the central and southern regions of the country, the traditional antipathy to livestock, coupled with a climate that made animal husbandry difficult, retarded any progress that might have occurred. The majority of farmers, small and large alike, preferred to avoid a 365-days-per-year commitment to the care of animals if they could get by with 60 to 70 days per year on their corn and beans, there being at that time no severe pressure of population and no great expectations of improvement in small operators' incomes. That the neglect of animals was not wholly a matter of climate can be seen in the small European colonies—of Germans, Mennonites, and others—where livestock was and is an important activity.

After the Mexican Revolution two new phenomena made their appearance, both of which might have been expected to give a stimulus to the livestock sector. One was agrarian reform, which over a period of sixty years has gradually broken up nearly all the estates and brought into existence million of small production units similar in size to the small family farms of Western Europe. The other influence was the demographic explosion, beginning effectively in the 1940s, which either caused farmers to divide their land among their several sons or created a large class of landless farm workers.

In practice the influence of each of these factors was attenuated. In order to safeguard the continuity of meat supplies to the urban population and exports of cattle to the U.S.A., the Mexican government in the thirties and forties granted to a number of large-scale ranchers what were called "inalienability certificates" which exempted their lands from subdivision for a period of 25 years. Meanwhile in the new ejidos to which land was massively transferred, the pastures were owned not by individuals but by the ejido collectively, so that no one had any responsibility or incentive to improve the grazing. And if an ejidatario acquired substantially more cattle than his neighbors, he might be denied the right to pasture them. Thus the new mini-farmers settled down to cultivating corn and beans, while livestock production remained for the most part in the hands of the reduced number of ranchers.

This created a paradoxical situation, exactly the opposite of that in Europe where mini-farms also predominated. The main field crops which, for cost reasons, ought to have been grown on a large scale, were being cultivated by mini-farmers, whereas livestock, which ought to have become the occupation of the mini-farmers, remained with the ranchers. Consequently the mini-farmers were and are condemned to a life of poverty, while the needed expansion of livestock production failed to take place because of insecurity of tenure and the agrarian legislation. Consequently Mexico became a large importer of milk products.

Nor was this all. Since approximately 1965 technical developments in the poultry industry have made it possible for hundreds of thousands of birds to be kept in a single production unit, thus achieving major reductions in the production costs of eggs and poultry meat. In other countries where this occurred, the small farmers were already well-established poultry keepers, and, although some of them were driven out of business by the superior efficiency of the new system, many others defended themselves successfully by forming cooperatives for the purchase of feed and day-old chicks and for marketing their output. In Mexico this innovation arrived before any significant number of small farmers were sufficiently established to maintain themselves in competition with large-scale operators. Today it is probably too late for the small producers to learn the techniques of the mass market. Similar technological innovations in hog production threaten also to deny that branch of production to the mini-farmers unless they quickly organize themselves.

In Mexico there is an extraordinary separation of the crop sector and the livestock sector, reflected in the language and the laws. In Spanish there exists no equivalent for the word "agriculture," which in English means the activity of producing food of all kinds. Neither is there any equivalent for the word "farmer." There is only an *"agricultor,"* who produces crops, or a *"ganadero,"* who manages livestock. Similarly in the Constitution of 1917, Article 27 talks about crop farms and livestock farms, laying down upper limits of size for each, but nowhere mentions or makes provision for mixed farms, an omission which, as we shall see, has done much harm. It is the same with the most recent Agrarian Reform Act of 1971, except that in one article it grudgingly permits a rancher to grow a limited amount of fodder crops, *provided that the output is solely for the use of his own animals.*

This divorce between crops and animals is continued in the bureaucracies and organizations. When one visits the local offices of the Department of Agriculture there will be much discussion of the local crops but no interest is shown in animal husbandry. When one contacts the farmers' organizations, there are the crop men and the cattlemen, operating different associations located in different parts of town. The ejidatario may own some animals, but he gets little help from the extension staff. The agricultural credit provided by the official bank is directed overwhelmingly toward crop production. The collection of statistics is focused 99 percent on crops, and the exiguous data assembled on livestock and livestock products can hardly be taken seriously. It is against this discouraging background that we have to undertake an analysis of Mexico's animal husbandry, its recent history, and its future potential.

CATTLE

The factual information in regard to cattle, as with all other elements of the livestock sector, is scarce and unreliable. It is impossible to reconcile the conflictive data published by the two agencies chiefly responsible, namely the Department of Agriculture (for annual statistics) and the Statistical Office (for census data). In the following sections we have used what, in each case, seem to be the least dubious figures, checking them when possible against other sources and against the experience of other similarly situated countries.

Numbers

In Mexico the cattle population appears to have increased by 36 percent between 1930 and 1950, and by a further 45 percent between 1950 and 1970 (census data); that is, it slightly more than doubled in forty years while the human population increased two-and-a-half times. Thus there were fewer animals in relation to people in 1970 than there were in 1930, and by 1980 this ratio had deteriorated further. As will be seen, this relatively slow growth in cattle numbers was probably compensated for by increases in cattle productivity, so that per capita consumption of beef and milk has been more or less maintained over the last few decades. But it certainly has not increased in the way one would normally expect under conditions of rising standards of living. By international standards, however, Mexico's performance has been satisfactory. Over the past thirty years its expansion in cattle has in percentage terms outdistanced all other Latin American countries and, indeed, all the advanced agricultural countries except Australia and New Zealand.

Certainly the cattle industry of Mexico is not a homogeneous entity. It should be considered in terms of its component groups, each adapted to the geophysical conditions of a specific region. To present a reasonably accurate but simplified picture, we may cite the three principal groups identified in a study by the Economic Commission for Latin America.[1] The first of these embraces the northern parts of the country where cattle are managed in the traditional ranching manner. The pastures have generally a low carrying capacity of 20 to 120 acres per head, according to district; the herds wander over great distances, and one man looks after on average 100 head. The young stock run with their mothers until the milk dries up, and there is virtually no commercial milk production. The product of the ranches is chiefly store cattle

[1]Comisión Económica para América Latina, *La industria de la carne de ganado bovino en México*, Fondo de Cultura Económica, México, 1975.

(steers) for export at an average weight of a little less than 400 pounds and also, but on a smaller scale, for sale to other Mexican cattlemen who fatten the animals for the domestic market. Most of the pastures are badly over-stocked, especially near the waterholes which are too few; many of the cattle are malnourished; birth rates are low; mortality is high due to malnutrition and disease; and only moderate technical progress has been achieved in recent years for reasons which will soon become apparent.

The second region is that of the humid tropics and subtropics, extending from Tamaulipas down to Chiapas and the peninsula. It is an area where natural pastures grow more abundantly and where in winter the "nortes" bring humidity which prevents the grasslands from drying out. In this region are located almost all of Mexico's artificial (sown) pastures, an innovation which began in the fifties, attained 10 million acres at the 1970 census and is destined to expand more widely. Because of its heat and humidity, however, the region is infested with cattle pests and diseases, especially ticks which debilitate the animals and reduce the commercial value of their hides. Yet it is the principal supplier of meat to the Mexico City market and is also becoming a major supplier of milk.

The third region, which comprises the Altiplano (apart from the north) and the Pacific coast states from Nayarit to Oaxaca, is much more heterogeneous and could be divided into several sub-regions. It contains the cattle-keeping of the mountain areas which is really mini-scale ranching, the rustic cattle of the plateau where the animals are kept as much for work purposes as for anything else, the islands of dairy production based on irri-gated alfalfa (in the Bajío) and the semitropical cattle farming of the Pacific coast where the tick infestation and other veterinary problems resemble those of the Gulf coast. This area provides milk and to a lesser extent meat for Mexico City and Guadalajara, especially from Guanajuato, Jalisco, and Michoacán. The Altiplano cattle are generally undernourished, and through the winter are maintained precariously on corn stalks and cactus.

In recent years cattle have multiplied most rapidly in the Gulf region, with increases of 50 percent or more in most states (Table 5.1). The northern regions have registered significant gains only in some irrigation districts and in Sonora, while the center regions record the least increase of any. In other words, the stock of cattle has increased mainly where pastures are rich and abundant or where alfalfa is grown with irrigation. It has expanded least in the central states in spite of good markets, which suggests the presence of serious physical obstacles.

Feedstuffs

The importance of the feed factor can be understood by relating the pace of expansion to the number of acres of pasture available per herbivorous animal.* Thus, except in Veracruz (high quality pasture), all the states in

---

*Cattle, sheep, goats, horses, mules, and asses have been converted to "livestock units." The relation is only rough because it cannot take into account the wide variations in grassland quality in the different zones.

TABLE 5.1

Livestock Numbers by Regions, 1978; and Livestock Annual Growth Rates, 1950 to 1980

Region	Cattle	Horses	Mules	Asses	Hogs	Sheep	Goats	Poultry
				thousand head				
Northwest	3,177	746	203	324	985	176	980	9,252
North	5,134	786	455	334	927	1,272	1,409	9,908
Northeast	1,432	224	131	133	315	410	828	8,517
North-center	2,495	838	585	309	1,108	1,923	1,430	6,774
West-center	6,165	1,151	451	864	3,915	891	1,448	38,226
Center	3,143	771	661	640	2,575	2,006	1,047	47,261
Gulf-south	7,732	1,735	663	588	2,495	1,152	943	26,969
Peninsula	640	195	58	41	259	21	19	2,963
Mexico	29,920	6,446	3,207	3,233	12,578	7,850	8,103	149,870
				annual growth rates				
Northwest:								
1950–70	1.9	0.2	– 0.8	– 0.3	0.9	– 2.6	3.8	6.4
1970–80	1.8	1.3	0.4	– 0.3	1.8	– 0.2	– 1.1	3.0
North:								
1950–70	1.4	0.2	– 0.5	0.7	0.9	– 2.9	0.2	1.8
1970–80	2.0	0.9	– 0.1	– 0.3	1.8	– 0.1	– 1.0	2.2
Northeast:								
1950–70	2.2	0.8	0.5	1.2	0.8	– 1.3	– 0.4	5.9
1970–80	2.0	1.0	0.4	– 0.4	1.8	– 0.2	– 1.1	2.9
North-center:								
1950–70	1.6	0.7	1.8	0.4	1.0	0.0	1.2	– 0.4
1970–80	1.6	0.9	– 0.3	– 0.3	1.9	– 0.2	– 1.1	2.9

TABLE 5.1
(continued)

Region	Cattle	Horses	Mules	Asses	Hogs	Sheep	Goats	Poultry
West-center:								
1950–70	1.2	1.8	1.7	− 0.6	1.0	− 1.2	− 0.6	3.2
1970–80	2.0	0.9	0.4	0.1	2.0	− 0.1	− 1.1	2.1
Center:								
1950–70	0.8	1.8	0.7	0.1	1.9	0.1	0.3	4.3
1970–80	2.0	0.9	0.5	− 0.9	3.5	0.2	− 1.1	3.3
Gulf-south:								
1950–70	3.5	2.3	3.8	1.6	3.2	2.6	1.8	2.5
1970–80	2.2	1.1	− 0.3	− 0.3	1.9	− 0.2	− 1.1	2.1
Peninsula								
1950–70	1.1	1.1	− 1.4	− 0.4	0.1	7.1	− 2.6	− 0.6
1970–80	2.2	0.9	− 0.1	− 0.1	1.8	0.2	− 1.3	2.3
Mexico								
1950–70	1.9	1.2	0.9	0.2	1.6	− 0.2	0.4	3.5
1970–80	2.0	1.0	0.1	− 0.3	2.2	− 0.1	− 1.1	2.7

SOURCES: Livestock numbers in 1978 from Department of Agriculture, Annual Livestock Report, 1979. Growth rates 1950–70 from farm censuses; for 1970–80 author's estimate on basis of livestock reports to 1978.

which cattle numbers increased (from 1950 to 1970) by 50 percent or more, had pasture coefficients of 5 acres or better per animal; whereas those in which the cattle increase was below 20 percent had already in 1950 a pasture coefficient of 2.5 acres or less. Thus, broadly speaking, the cattle population increased least where there was already evidence of overstocking and most in areas where the grazing had not yet been fully exploited. Expressed in this way the statement is indeed a platitude, but the highlighting of this aspect of the problem has implications when we come to examine the potential for future expansion.

From the historical evidence there emerges a pronounced trend for the pasture coefficient to decline. Except in the northwest and northeast regions, this has occurred all over the country and most conspicuously in the center region. In no less than twelve states, mostly of the central plateau, the coefficient by 1970 had fallen to less than 2.5 acres per animal unit, and in two to less than one acre. By 1980, although we have no pasture data, the situation was almost certainly worse.

In the early seventies the Mexican government set up a technical commission to study and report on the carrying capacity of the nation's grasslands, and the commission established separate coefficients for each of a number of zones in each of the 32 states. In almost all zones and regions the commission's figures were higher than the prevailing rates, which means that overstocking was widespread.

Another more menacing aspect to this problem concerns the differences between the private farmers on the one hand and the ejido and community cultivators on the other. For instance, in 1970 the average area per herbivorous animal unit for the private farm sector was 10 acres, while for the ejidos and communities it was less than 5 acres. The gap between the two groups has narrowed over the past thirty years, but it remains considerable. Again in the central region the situation is the most acute: the state of México recorded only 0.89 acres of pasture per animal unit in ejidos and communities, while Tlaxcala reported only 0.37 acres.

TABLE 5.2

**GRASSLAND ACRES PER HERBIVOROUS ANIMAL UNIT:
SELECTED STATES, 1970**

(acres per animal unit)

State	Private farms over 12.5 acres	Ejidos and Communities
Sonora	17.8	13.3
Durango	15.1	8.2
Michoacán	5.1	2.1
Puebla	4.0	1.5
Veracruz	3.0	1.5
México (state of)	1.4	0.9
Tlaxcala	2.1	0.4

SOURCE: Statistical Office, farm census of 1970.

NOTE: Grassland includes areas of cultivated grass.
An animal unit = one each of cattle, horses, mules, and asses; 10 sheep and 10 goats.

Yet another interesting observation is that whereas the ejidos' ratio has varied little over recent decades, the private farmers' ratio has fallen significantly. What seems to have happened is that as a result of the massive transfer of land, especially of pasture, from the private to the ejido sector, the private farmers have tried to defend themselves by increasing their cattle numbers which involved maintaining more animals on fewer acres.

More disturbing is that an important proportion of the limited supply of grass is being consumed by non-productive animals. (For this purpose "non-productive animals" include horses, mules, asses, and oxen which chiefly perform field work and provide transport.) For the country as a whole these non-productive animals constitute nearly one third of all herbivorous animal units, but the situation varies greatly from state to state. In Baja California Norte and in Sonora the proportion is around 10 percent, whereas in Puebla it is over 50 percent and in Tlaxcala over 70 percent. Generally the states with the highest percentages are those in which least technical progress has occurred.

Again, in this respect, the situation is worse in the ejidos and communities. In Sonora the non-productive animal percentage is 25.5 percent for these classes of operators against 7.7 percent for private farmers; in Puebla the respective figures are 65.8 percent and 21.7 percent and in Tlaxcala 75.3 percent and 36.3 percent (Table 5.3). Another significant indicator is the proportion of the cattle population devoted principally to work purposes. This proportion reaches 33 percent in ejidos and communities in the states of México and Tlaxcala and even attains 48.6 percent in Puebla, whereas on private farms even in these states the percentage is negligible. Of course, sooner or later the situation may be expected to change, by progressively substituting mechanical power for the present excessive animal power. Furthermore, it is probable that, even without the introduction of more tractors and vehicles, the farm operations could be performed by fewer horses, mules, etc., many of which are kept primarily for prestige purposes.

TABLE 5.3

UNPRODUCTIVE ANIMALS AND OXEN:
RELATIVE IMPORTANCE BY TYPE OF FARM, 1970

State	Unproductive Animals as % of All Herbivorous Animal Units		Oxen and Working Cows as % of All Cattle	
	Private Farms over 12.5 Acres	Ejidos and Communities	Private Farms over 12.5 Acres	Ejidos and Communities
Sonora	7.7	25.5	0.1	0.2
Durango	19.3	44.6	2.1	1.8
Michoacán	18.4	47.9	7.6	17.9
Puebla	21.7	65.8	6.5	48.6
Veracruz	8.2	46.7	1.4	13.0
México (state of)	21.3	51.7	8.6	32.7
Tlaxcala	36.3	75.3	5.1	33.2

SOURCE: Statistical Office, farm census of 1970.

NOTE: Unproductive animals refer to horses, mules, and asses.

Thus we have to face the discouraging picture that not only is an important part of the nation's pasture resource being used by non-productive animals, but also a significant proportion of the nation's cattle is being used for work rather than for food production. In both respects the position is more disadvantageous in ejidos and communities than in private farms,yet this merely reflects the technical backwardness of these groups and that they comprise a far higher proportion of mini-farmers.

This persistent retention of more work and prestige animals than are really needed is a characteristic of all agricultural economies at a certain stage of their development. It was, for instance, a major problem in Europe until World War II. It is only when subsistence farming is almost completely replaced by commercial farming, and when farmers have become cost-conscious and begin to manage their inputs in a business manner, that the superfluous animals can be expected to start disappearing. Mexico may hope to begin this transformation toward the end of the period under consideration here.

These obstacles to a rational evolution of the cattle industry would not be so serious if the country possessed ample supplies of other feedstuffs, in particular feed grains and oilcake, but that is not the case. During the past few years cereals and oilseeds have had to be imported on a substantial scale mainly to satisfy human requirements but also to a considerable extent for use as animal feed.

Moreover the prices which the government guarantees (or fixes) for basic feedstuffs on the one hand and for livestock products on the other, do not relate to one another in a manner which would encourage the extensive use of balanced feeds in cattle fattening or in dairying. For example, in the U.S.A. the traditional ratio between the price of corn and the price of carcass beef has been 1:20 or even 1:24, whereas in Mexico over the years the ratio has averaged 1:10. Where such a poor price relationship persists it is not generally economically sound to undertake feedlot operations based on cereals and oilcake. Similar considerations apply to the economics of milk production, which is why the only intensive dairying activity in Mexico is located in districts where alfalfa supplies are plentiful. We shall be discussing presently how this policy toward the livestock industry might be modified; for the moment let us note the real difficulties which exist.

Beef Production

Although many difficulties are encountered in trying to arrive at a credible estimate of the volume of Mexico's beef production, we will begin with the more reliable component, namely the exports of live cattle and of beef. Before World War II there had been an active trade in live cattle to the U.S.A., but this collapsed when foot-and-mouth disease ravaged the Mexican cattle herds in the late forties. In the fifties and sixties it steadily recovered until by the late seventies the export attained over 700,000 head annually, equivalent to some 56,000 tons of carcass meat at the time of export. During the same time an export trade in beef was also being developed which by 1979 attained over 30,000 tons per year.

The great bulk of the exports originates from the northern states of Sonora, Chihuahua, and Coahuila, although when prices are high in the U.S. market, cattle may be exported from Durango and even more distant states. If, as a hypothesis, the total live cattle exports are expressed as a percentage of the combined cattle populations of Sonora, Chihuahua, and Coahuila, then this percentage rose from 13 percent in the early sixties to 19 percent in the seventies. Unfortunately we cannot say whether this apparent increase in offtake represented a genuine increase in the productivity of the herds in those districts, or whether it merely constituted a diversion to export of animals which formerly would have been slaughtered for the domestic market. This latter explanation is plausible because we know that during the same period beef production for home consumption was expanding strongly in the Gulf states.

To the question of what has been the total volume of production, both for export and domestic markets, all Mexican authorities admit the published figures are far below reality. This is explained partly because a considerable amount of the slaughtering takes place outside the officially registered slaughterhouses, and partly because even the slaughterhouse statistics are believed to be underestimates for tax evasion reasons. We can arrive at a more reliable figure by using two approaches: (1) by comparing beef production per thousand head of cattle in various Latin American countries; and (2) by comparing per capita beef consumption in these countries. By both yardsticks it becomes apparent that the official production figures should be almost doubled. For instance, in respect to per capita consumption, the official figure of domestic beef supply would represent 35 pounds, which is less than in all but two of the Latin American countries. Our figure would be around 68 pounds, which would place Mexico in a middle range among these countries.

We are thus led to believe that in the late seventies total beef production (for domestic needs and for export) somewhat exceeded 1.2 million tons per year (carcass weight). In order to establish a historical series (Table 5.4), we further assumed a one percent per annum growth in the productivity of the herds. With this coefficient and with the trend in herd size as reported by the censuses, it was possible to calculate backwards: for example, the output in 1950 would appear to have been just over 500,000 tons. The reasonableness of a one percent productivity growth rate can be checked as follows. If we were to take a rate significantly lower than one percent, then the output of livestock products in the earlier years would have greatly exceeded in value the output of the crop sector—a relationship which no one in Mexico would be disposed to believe. On the other hand, if we were to assume a productivity growth rate of 1.5 percent or 2 percent, we should encounter an insoluble difficulty in pasture productivity, a topic which in any case must at this point be mentioned.

TABLE 5.4
ESTIMATES OF MEAT PRODUCTION AND CONSUMPTION: 1930 TO 1980

Year	Annual Production of Meat (thousand tons carcass weight)					Utilization (thousand tons carcass weight)		Consumption per Capita (pounds)
	Beef	Pork	Mutton & Goat	Poultry	Total	Export	Domestic	
1930	269	86	38	32	425	10	415	55.3
1940	342	131	47	58	578	16	562	63.1
1950	512	196	62	101	871	6	865	73.9
1960	636	188	76	125	1,025	50	975	59.5
1970	906	328	79	296	1,609	100	1,509	65.7
1980(p)	1,217	448	84	426	2,175	55	2,120	68.4

SOURCE: Author's estimates, see Appendix.

NOTE: Exports include carcass weight equivalent of live animals. (p) refers to preliminary data.

Earlier in this chapter several references were made to the overstocking of pastures and their consequent deterioration, especially in the northern regions. However, we now have the problem of reconciling this statement with the fact that the output of meat per acre of grassland has been increasing—indeed has increased quite rapidly. Thus, between the years 1950 and 1970 the total pasture area increased 16.5 percent while the output of meat from cattle, sheep, and goats increased by 71.6 percent. This implies an increase in pasture productivity of 47.3 percent over twenty years.

At least two objections can be raised against this procedure and its conclusion. First we have included the "cultivated grasslands," which appeared as a separate item in 1970, as equivalent to ordinary pasture in feed value per acre, which of course is not true. How much superior they are varies from district to district: maybe tenfold, maybe one and a half times. If we assume an average of threefold then a correction for this factor increases the number of pasture "units," and reduces the rise in pasture productivity from 47.3 to 38.4 percent.

The second objection relates to the contributing of cereals and oilcake to the production of meat. If this contribution had been increasing very rapidly, it might largely eliminate the need to assume higher pasture productivity. Now in Mexico the manufacture of animal feeds has certainly been a rapidly expanding industry, but much of its output is sold to hog and poultry producers. Cattle fattening based on these feeds has been expanding only slowly because of the unfavorable price ratios mentioned earlier. One may be prepared to accept that by the seventies the contribution of concentrates to total cattle feeding had become *somewhat* higher but not *much* higher; and if this were so it might reduce our pasture productivity increase from 38.4 percent to say around 30 percent. That would still leave us with a phenomenon which is difficult to reconcile with the agronomists' contention of overstocking.

Of course, this does not add up to saying that the agronomists are mistaken. The fact of overstocking has been carefully documented in several studies, but these have referred to particular zones almost all of which are in the northern or in some of the central states. What our analysis does is to warn against generalizations applied to the whole Republic. Overstocking in certain districts occurs concurrently with understocking in others; declines in pasture productivity occur in some and simultaneously productivity increases occur in others. What seems to have happened is that more intensive utilization of the newer grassland areas of the Gulf and south has more than offset pasture deterioration in other parts of the country. However, this favorable trend may be coming to an end as the newer areas attain full exploitation.

Milk Production

Information on the volume of milk production is as contradictory as that on beef production, the main difference being the relative positions of the agencies regarding the reliability of their published data. In respect to milk, for unexplainable reasons, the census data appear sensible whereas those of the Department of Agriculture seem far too low. By extrapolating from the last census and taking account of some sensible figures published by the National Institute of Nutrition, we arrive at an estimate for 1980 of a little over 6.4 million tons as the production of cows' milk plus about 200 thousand representing goats' milk.

To help us establish a historical series of milk production, we have data on the number of cows but need to make an assumption about the trend in average yield per cow. From the census data and other evidence it would appear that prior to about 1960 the yield per cow was rising at about 1.5 percent per annum overall, while since 1960 the increase has been closer to one percent. On these assumptions the output may have been some 2,550 thousand tons in 1950, 3,670 thousand in 1960 and 4,700 thousand in 1970 (Table 5.5).

These estimates would imply a per capita production of Mexican milk rising rapidly to about 108 quarts in 1960, since when there has been a notable decline.* In short, milk production over the last twenty years has not quite kept pace with the growth in population. The explanation must be sought in a combination of technical, economic, and institutional factors to be examined later.

In Mexico, as in many other countries where ranching is of importance, a significant proportion of the cow population is not regularly milked; the cows are left on open range to suckle their calves until the milk dries up. In such circumstances, if one were to relate the estimated national production of milk to the reported number of adult female cattle, one would arrive at a meaningless figure. What would be rational would be to relate the output to the number of cows being milked, but since in Mexico no one knows what this number is, the calculation cannot be made. Various authors have mentioned figures of average milk yields in the range of 950 to 1,150 quarts per cow, but these are mere surmises.

In each census since 1940 the implied milk yield (relating total milk output to total cows) has been substantially lower in ejidos and communities than on private farms. No one believes that the role of ranching (and therefore of unmilked cows) is higher in ejidos than on private farms, but rather the reverse. Therefore, the difference in yield, even calculated in this unsatisfactory manner, almost certainly reflects a true difference which the census data probably underestimate. Now it happens that since 1940, because of the creation of many new ejidos, the number of cows reported in ejidos has nearly trebled, while the number on private farms has increased by only 56 percent. That low productivity cows are increasing in numbers more rapidly than high productivity ones itself constitutes part of the explanation of the slow growth in milk production.

---

*Total per capita consumption has not declined because it has been assisted by substantial imports of powdered and condensed milk (see Table 5.5).

TABLE 5.5

PRODUCTION AND CONSUMPTION OF MILK AND EGGS: 1930 TO 1980

Year	Cow and Goat Milk				Eggs	
	Production (thousand tons)	Import (thousand tons) tons)	Consumption (thousand tons)	Consumption per Capita (U.S. quarts)	Production (millions)	Consumption per Capita (eggs)
1930	1,227	—	1,227	78.3	788	48
1940	1,681	3	1,684	90.6	1,451	74
1950	2,551	36	2,587	106.0	2,513	97
1960	3,669	11	3,680	107.9	3,099	86
1970	4,694	370	5,064	105.6	7,376	145
1980(p)	6,680	1,000	7,680	113.0	10,620	156

SOURCE: Author's estimates, see Appendix.

NOTE: Imports of milk products converted to fluid milk equivalent. No significant external trade in eggs. (p) refers to preliminary data.

OTHER LIVESTOCK

Sheep

Sheep farming, whether for wool or for meat production, has never achieved much importance in Mexico. The consumption of mutton and lamb has never become a national habit, while the wool textile industry's dependence on imported wools has been increasing.

The sheep population of the Republic (according to the census) was slightly lower in 1970 than in 1950, just below 5 million. According to the Department of Agriculture it rose rapidly during this period and in 1970 attained 7.9 million, and has stabilized since then. No one has yet succeeded in reconciling these divergences. Sheep have a certain importance in the semiarid regions of the north, but the largest concentrations are found in Hidalgo, México, Puebla, and Oaxaca. A₋ one would expect, sheep are favored mainly in the mountain zones of the Altiplano, but since grazing in these areas is almost everywhere poor, and since the land tenure is so fragmented, it has never become possible to develop sheep farming on the scale which would make it profitable.

Another important consideration is the climate of the Altiplano, with its seven months of winter drought and five months of summer rains. We can find considerable areas of grassland which could carry a substantial sheep population in the rainy months but which do not do so because there would be nothing to feed them on in the dry months. The impossibility of shifting the flocks from one zone to another according to the season constitutes a serious defect in the prevailing institutional arrangements to which we shall return later in this chapter.

Goats

Professional opinion is divided on the subject of goats. On the one hand are the environmentalists who regard goats as a menace to the natural vegetation, as a major cause of soil erosion, and as something to be extirpated. On the other are the agronomic economists who maintain that a herd of goats may be the principal source of income to many farm people who have no alternatives, and that either this herd survives in areas where the vegetation is already below the level of acceptability by any other class of animal, or it can be organized to respect the environment as well as any other domesticated animal, provided that the herd be well-fed and well-managed. The fact is that goats still remain a mainstay of income to many farms in marginal areas in Mexico, for example in Coahuila and Oaxaca.

According to the census the goat population increased from 6.5 to 8.5 million between 1930 and 1950, but then only to 9.2 million in the next· twenty years. According to the Department of Agriculture it reached 8.9 million already in 1960 and has been falling gradually since. Despite these conflicts, both sources are agreed that Coahuila is the principal goat state, followed in importance by San Luis Potosí, Zacatecas, Puebla and Oaxaca. There are areas in the northern sierras of Cóahuila where goats have long been the basis of the rural economy, and where goats' milk fetches a higher price per liter than cows' milk, whether the former be used for direct human consumption or for cheese making. It may be noted in passing that in some western European countries consumer demand for goat cheeses has grown so much that it has become difficult for the cheese makers to obtain goats' milk in sufficient quantities.

Attempts have been made to introduce into Guerrero the breeds of goats which have succeeded in Coahuila, but the experience has been discouraging. Breeding stock from Saltillo has been a failure in Altamirano. On the other hand, there could probably be a useful expansion of goat keeping in the dry zones of San Luis Potosí where the ecology is sufficiently similar to that of Coahuila.

Hogs

Mexico's hog population experiences vicissitudes occasioned partly by the hog cycle common to all countries and partly by more general factors. The censuses report an increase in numbers of 87 percent between 1930 and 1950, followed by a further increase of 37 percent between 1950 and 1970 to reach 9.5 million. The Department of Agriculture publishes much higher figures, though its rate of increase is slow, and arrives at 11.6 million in the 1970s. The hog population is widely dispersed throughout the Republic and features prominently in the unpaved alleys of the villages. In terms of value of output, hogs are four times as important to Mexico's farmers as sheep and goats combined, and have only recently been surpassed by eggs and poultry meat.

Very little of the national output has as yet been concentrated in large production units. One major reason is that this method of pork production depends almost entirely on cereal supplies, and as long as the relative prices of feed grains and hog meat remain unfavorable, production cannot be successfully undertaken in specialized enterprises. To be profitable, hog fattening has to incorporate into the ration a substantial amount of farmyard waste which costs the hog farmer virtually nothing.

Hogs remain a sideline activity for a million small-scale operators. In a survey undertaken in 1973 covering farms which possessed more than ten hogs each,[2] just over two-fifths of the farms had hog production as their main activity, another two-fifths had crops and/or other livestock products as their main interest, while the remainder had a nonagricultural main activity. This suggests that hog keeping is still largely in the hands of general farmers and nonfarmers.

The survey also revealed the low technical level of operations. The sows averaged only 5.4 deliveries in their lifetime, the average litter was 9.5 of which only 7.5 were weaned—a high mortality rate. An additional three percent mortality occurred during fattening. Altogether the extraction rate, i.e., hogs slaughtered per 100 of hog population, was 54 percent, compared with 170 percent in Denmark. Diseases and parasites were the chief cause of

---

[2]Banco Nacional Agropecuario, *El mercado de ganado porcino en México,* México, 1973.

losses. The hogs required an average of 165 days fattening and were sold at 230 pounds liveweight, a very slow rate of gain.

The use of feed mixtures has expanded rapidly over recent years, the principal ingredients being corn, sorghum, chickpeas, alfalfa, and various oilseed meals, a part of these last being imported. The prices of feed mixtures have been rising more rapidly than the prices of hogs so that producers' profit margins have been squeezed. And since 80 percent of the producers sell individually rather than through cooperatives or other groups, they are ill equipped to defend themselves.

The volume of production, as with other meats, has been a subject of conflicting estimates. We believe that annual production in the late seventies has been about 450,000 tons, which represents some 13 pounds per capita. This is a low level of consumption compared with the U.S.A. at 64 pounds, Poland 88, and the German Democratic Republic 97 pounds.

Strangely enough, while in most other countries the consumption of pork is concentrated in the lower income groups, in Mexico the reverse appears to be the case. A study undertaken by the Department of Industry and Commerce indicated that among urban consumers the importance of pork was higher in the higher income groups, almost attaining parity with beef where incomes exceeded 3,000 pesos per family per month. This occurs because in Mexico the prices of pork and bacon tend to be somewhat higher than the price of beef, the opposite of what prevails in the U.S.A. and Europe. However, if during the next few years consumers' incomes in Mexico start improving rapidly (because of oil revenues), a significant increase in pork consumption may be expected.

Poultry

This has for many years been the most dynamic section of Mexico's livestock industry, partly because of important technological innovations not paralleled in respect to other categories of livestock. The most vital of these was achievement of a capacity to control effectively the major poultry diseases, which thus reduced to an acceptable level the risk of keeping large numbers of birds under a single roof. This, coupled with systems for the automatic supply of feed and of drinking water, enormously reduced manpower requirements and hence the unit costs of production of poultry meat and eggs.

As a result there emerged mammoth enterprises with as many as half a million birds in a single unit. For the consumer the consequence was that poultry meat, having been the most expensive of all the meats, now became almost the cheapest, and that the real price of eggs fell sensationally. In forty years the importance of poultry meat has risen from 10 percent to 20 percent of total meat consumption in Mexico, while the per capita consumption of eggs has doubled.

The poultry population is believed to be around 150 million in 1980, having nearly trebled in thirty years. Large commercial production units are located in Sonora, around Monterrey and in the Mexico City area. Estimates of poultry meat production vary widely; our own guess is that by 1980 it reached some 426,000 tons per year. Even greater uncertainty attaches to egg production, partly because small farm output remains important. A fair guess might be around 10,600 million eggs in 1980. It is impossible to evaluate the efficiency of egg production by the yardstick of number of eggs produced per laying bird, because (as in the case of cows) we do not know what proportion of the total poultry stock is devoted to the meat side and what proportion to eggs. One may reasonably assume that the most technically competent units approach international standards in their efficiency.

The expansion of the poultry industry, as that of the hog industry, has received an enormous impetus from the arrival of sorghums as a commercial crop in Mexico, and its expansion up to a level of five million tons in 1980. Had it not been for this development, either massive feedgrain imports would have been necessary (even so, certain quantities are imported), or alternatively the hog and poultry businesses could not have expanded so fast. The sorghum production has been supplemented by increasingly large quantities of oilcrop residues, especially of soya and safflower (which are also new crops), thus enabling the feeds concentrates industry to expand into a major economic activity. Whether in the years ahead Mexico's agriculture can continue to provide the quantities of feedstuffs which will be required in the livestock sector to meet the ever-growing demands of the consuming public, is the question addressed in the following section.

LIVESTOCK PRODUCTION TARGETS

When we reflect on the difficulties the livestock sector has encountered over the past thirty years, its performance has been remarkably good. The output of meat and milk has more or less kept pace with the growth in population while the poultry industry has grown even faster, and all this in spite of official neglect bordering on hostility in recent years.

But looking to the future there are serious reasons for being alarmed. Population is still increasing rapidly, per capita purchasing power will be on an upward trend, and consumers will be diversifying their diets. In Chapter 2 we decided to work on two alternative assumptions of income growth to the end of the eighties, and accordingly two rates of demand growth were established for each of the principal food stuffs. The lower rates were 3.5 percent for milk, four percent for meat, and five percent for eggs; while the higher ones were four, five, and six percent respectively.

Remembering that population is expected to be growing at 2.7 percent, we see the slow growth alternative implies only a modest increase in per capita milk and meat consumption, while only in eggs would any significant consumption increase occur. This would be a gloomy prospect, and it would surely be yielding to undue pessimism to formulate production targets on this basis; or at least the pessimist's growth model should be kept in reserve until the implications of the more optimistic alternative have been examined. It is not so wildly optimistic, however, to hope that per capita consumption of milk might increase at somewhat more than one percent per year from its present low level, and of meat at over two percent.

We will therefore work with the higher growth rates of demand and must now translate these into production targets. In any planning for the livestock industry it is necessary to take a long view, because, especially in the case of beef and dairy cattle, the size of herds and their output cannot be rapidly increased from one year to the next. Moreover, in Mexico the livestock dilemma seems likely to persist for many years ahead, as we shall presently see.

Thus a projection model for the eighties should be regarded as just one installment of a longer-term development. For simplicity we will assume that exports of steers and beef and imports of milk products will in 1990 bear the same relation to domestic consumption as in 1980. In other words, in our model national production should be growing at the same rate as demand; and, as indicated, we will use the higher of the two demand alternatives.

The expansion of production is, of course, composed of two elements: improvements in average output per animal and increases in the number of animals. As regards the former, it was estimated earlier that in recent times the productivity has been increasing at about one percent annually in meat and milk, and at about two percent in poultry meat and eggs. It seems reasonable to suppose that in all livestock classes except poultry the one percent rate can be maintained in the coming years. However, in the poultry industry the technical revolution has already been accomplished, so that it is difficult to imagine any sensational further gains. Therefore a rate of increase of only 0.5 percent per annum is assumed for the future productivity of poultry.

With this productivity assumption, it can be deduced that to meet the demand targets animal numbers would have to grow at the following annual rates during the eighties: beef cattle and hogs four percent, dairy cattle three percent, sheep and goats 0.5 percent and poultry 5.5 percent. These are in themselves formidable growth rates. Moreover, they involve almost doubling (except in poultry) the best growth rates achieved in any previous decade.

We next have to ask how these animals are going to be fed. It is not a question of merely supposing that feed supplies should grow at the same rate as animal numbers—they will need to grow faster. Since the model has assumed an increase in productivity per animal, most of this has to come from giving the animals more and better food; only a small fraction can come from improving the animals' physical capacity to convert feed into meat, milk, etc. Therefore we must reckon that feed supplies will need to be expanded by almost the same amount as the *output* of livestock products which comes to 60 percent in the decade.

It is convenient to consider separately the feed derived (a) from pastures and (b) from annual crops, and to take as a first assumption that each should be expanded to the same extent. Considering first the contribution required from grasslands, we suggested earlier that between 1950 and 1970 pasture productivity probably increased by about 30 percent and, in view of what the agronomists report on pasture deterioration, we cannot expect so large an increase in the coming years. However, we have noted a continuing process of conversion of some areas to cultivated grassland in zones of good rainfall, so perhaps it may be legitimate to retain 15 percent as a projection for this ten year period. The remainder of the growth in grass supply, i.e. 45 percent, would have to come by expanding the pasture area from 194 million acres (assuming the 1980 figure approximates the 1970 one which is probably the case) to 281-million acres, an enormous change for such a short period.

What then would be a realistic prospect for pasture expansion? This is in part a theoretical question, because the answer would depend on how far the nation was prepared to sacrifice other valuable assets, in this case part of its productive forests. Leaving aside such a drastic change in land use, certain estimates were put forward in Chapter 3 which give an indication of what might be feasible. It was suggested that the opening of new land in the southeast might provide some 7.5 million acres of grassland; in addition rather more than one million acres might be secured through drainage projects under the National Water Plan. In other parts of the Republic perhaps another 3.5 million acres could be found, making some 12 million acres in all. But these 12 million could not all be won within a mere ten years; such a program would extend to the end of the century. Maybe seven million could be counted on for the eighties, bringing the total to 201 million, which is still a far cry from the 281 million needed.

Turning now to the feeds required from annual crops, and considering first the likely increase in per acre yields, we may estimate that a two percent per year improvement is the maximum that can be optimistically expected. This would provide 22 percent more feed out of the 60 percent needed during the decade. It may further be estimated from the evidence of the seventies and from our enquiries in Chapter 3 that the harvested area of feed crops might be increased at about the same pace as crops in general, namely one percent per year. These two factors together give a feed from crops increase of 34 percent during the eighties.

Thus, taking the grass and feed crops in combination, we are left with an enormous shortfall in supplies. Because no reliable data exist on the relative importance of these two classes of feed it is a little difficult to quantify the size of this gap. However, it can be roughly estimated that each provides about half the total.* On this basis we can evaluate the problem which is going to face Mexico. For pasture we have assumed a 15 percent increase in productivity plus a three percent increase in area, while for annual crops the increases are 22 percent and 10 percent respectively. The model will therefore look like this:

	1980	1990
	(in hypothetical units)	
Grass	50	59
Feed crops	50	67
Total Supply	100	126
Requirements		160
Deficit		34

---

*Approximately 40 percent of the total output of livestock products is derived from hogs and poultry which eat only annual crops. In addition certain annual crop products are fed to cattle, notably alfalfa and some concentrates, which may bring the total crop contribution to around 50 percent, the remainder being derived from the grasslands.

Two important features of this table deserve comment. First, the proportion of grass in the total supply may be expected to fall. This means that a larger proportion than hitherto of the nation's beef supply will have to come from feedlots and more of the milk from stabled cows fed on concentrates. (And all this would require price changes.)

The second feature is the large overall deficit, which signifies that at the end of our period Mexico would be importing about half its (nongrass) feed supply if the livestock production targets are to be met. Many Mexicans are unwilling to contemplate this possibility. Some of them assert that the nation should not try to imitate the consumption habits of the rich countries (though it is doubtful whether, when oil makes Mexico rich, consumption changes could be prevented except by dictatorial action). Others seek to put forward alternative figures to make the model look less startling, but no reasonable modifications in the figures would appreciably alter the order of magnitude of the feed gap.

However, for the purposes of the remainder of this chapter it is better to assume that the nation wishes to expand its levels of food consumption, and that we should seek for policies which would make this possible. If the policies failed, the nation would be condemned to a long period of poverty anyway, whereas if agricultural modernization succeeds, the cost in terms of feed import requirements might prove to be somewhat less than envisaged in our model.

IMPLEMENTING THE TARGETS

Let us therefore examine what steps may be necessary to accelerate the modernization, the productivity, and above all the output of the country's livestock sector. Any such exercise in policy formulation involves three main components, each to be considered in turn: the technical, the economic, and the institutional.

Technical Improvement

Always in the livestock sector, whatever the country, the technical problems fall into three basic groups: breeding policy, the control and eradication of diseases and pests, and finally the improvement of animal nutrition through a more copious and diversified feed supply. In each of these fields there is the research aspect and the dissemination of knowledge aspect. Since some of these matters will be touched on in other chapters, only those essential to the formulation of livestock policies will be mentioned here.

With breeding policy in respect to cattle, the most encouraging feature has been the persistent search by the ranchers of the north for breeds and crosses most suited to the environments in which they operate. In pursuit of this objective many have settled for Herefords, many for Zebu, some for Holstein or Aberdeen Angus or Charolais, or for first crosses of one of these with Zebu. This variety of breeds indicates partly the variety of ecological conditions in the nothern states and partly that most of the cattlemen have proceeded by trial and error, in the absence of any systematic research activities by the government.

Moreover, a considerable proportion, perhaps one third, of all the cattle in the north is "criollo," i.e., unimproved indigenous mongrels. The government should be developing more ample breeding research programs for these regions, having in view two aims: first the fixing of the most appropriate breeds for producing steers for the American market and for fattening in other parts of Mexico, and secondly the fixing of breeds for the dairy production in the Laguna and around Monterrey.

In the Gulf area one finds few big ranchers but rather small herds operated by ejidatarios or private farmers. Consequently, private initiatives in breeding are rare while the few government research stations lack any settled breeding policy. Much depends on the personal preferences of the technical director at the moment: one may aim to establish a beef breed and a milk breed, while another is convinced that dual-purpose animals should be the goal. In the central and western regions, apart from a few small isolated efforts, one finds no breeding research whatsoever.

These deficiencies stem from the authorities' basic lack of interest in animal husbandry and from their inability to visualize the grave livestock dilemma of the future. It is already clear that meat and milk production will continue to expand in the Gulf and southeast, especially if new lands are opened, and that some western regions will also make a larger contribution. It is therefore imperative to set up one or more research stations in each of these regions, each station with a well-defined breeding program. Probably each would have a beef herd, a dairy herd, and a dual-purpose herd (utilizing distinct breeds), in order to measure their relative performance over, say, a ten year period. Thereafter the extensionists could begin to make useful recommendations.

Breeding programs in respect to other classes of livestock are conspicuous by their absence. A few specialized hog producers have evolved breeding policies out of their own practical experience; but there remain as many hog breeds as there are states in the Republic, and the majority of hogs are of unknown parentage. A similar situation prevails in regard to sheep and goats, neither of which is considered of sufficient economic importance to justify research expenditure. Yet in large areas of the country the pastures are so poor that only these animals can graze them. It also seems likely that a market for mutton could be developed, and goat meat is one of the most expensive delicacies when it could be one of the cheapest meats. Both goat and sheep's milk, either fresh or in the form of cheese and yogurt, could be much more widely popularized.

To exploit these opportunities would, however, require the initiation of serious long-term programs in sheep and goat breeding, with breeds adapted to each end-product. If this were done, these activities could become more economic than they are now, hitherto unused (because poor) pastures could be mobilized for production, and the incomes of marginal farmers in remote areas could be greatly improved.

The problem of animal pests and diseases is much less one of research than of the more widespread adoption of the necessary remedial and preventive measures. Diseases and pests continue to be the principal cause of the high mortality among calves; they make the animals inefficient converters of the feed they consume; and they reduce the market quality of the meat and of the hides. Expenditure on control and eradication would have one of the highest rates of return of any investment in the livestock sector.

Recently the government has considerably amplified its allocation of funds for these programs and has been training a larger number of veterinarians. Yet still there is a shortage, still there are states which have only one qualified veterinarian (together with three or four hard-working but only partially trained assistants) to serve the entire state. The Rural Credit Bank makes funds available for the construction of dip tanks for tick control, yet still extensive districts lack this facility and large numbers of farmers do not appreciate the need for control measures.

Finally we come to the supply of animal feedstuffs. From our rather detailed analysis of this topic it appears that a grievous shortage is likely to emerge in the coming years. Nevertheless, it is worth asking whether it might not be feasible to initiate some programs which would mitigate the severity of the problem. For instance, the adaptation of grass varieties to arid and semiarid regions and to the humid tropics has long been studied in a number of countries. Much might be achieved by establishing research units in key zones, obtaining expert advice from abroad and setting up trials with exogenous varieties from Australia and elsewhere. In respect to the much neglected pastures of the northern ejidos, financial incentives could stimulate the digging of more waterholes, fencing for rotational grazing, building soil-retention banks, etc.

We have mentioned the frequent cases of pastures whose carrying capacity is limited by the number of animals they can sustain through the dry season. This capacity could be augmented in two ways. First, cattle ranchers should be permitted to cultivate larger areas of fodder crops without the risk of losing the legal security of their lands. Secondly, arrangements could be made between pairs of ejidos (or private farms), one in a pasture area and the other in a crop growing area, so that for the dry season the herds could be moved from the former to the latter to feed on crop residues, such as has for centuries been the custom in Mediterranean countries.

In respect to feed concentrates, it is now technically feasible to look beyond mere feed grains and oilseed crops and their modernization. Although the commercial utilization of algae is yet in its early stages, petroleum protein is already being manufactured in significant quantities for animal feed in France and in Italy, and even at present petroleum prices this emerges cheaper than the protein in vegetable oilseeds. Mexico has a rapidly expanding petroleum industry and by using less than one percent of her production she could provide all the protein requirements of her livestock population during the coming decades. In fact a project for constructing a pilot plant is in the planning stage as of this writing.

Economic Incentives

References have been made to the discouraging trends over recent years in the prices of livestock products. Of the four chief products, only beef shows a steady rise in real prices over the last twenty-five years, pork and milk prices have remained relatively stable, while egg prices have fallen substantially (see Fig. 9.2). The remarkable feature is that in spite of these not particularly encouraging price movements, the volume of production of all four commodities has registered a continuing increase. This suggests that production costs were declining to such an extent that a modest expansion of production was profitable, and of course the outstanding example is that of eggs where the technical revolution in production methods was so great.

It would require a detailed analysis of the structure of costs for each of the different products and in different regions in order to ascertain what has been happening. We do know, from an index published by the Department of Agriculture, that the price of feed concentrates has been rising in real terms over the past twenty years. We also know that a large number of crop products have experienced falling production costs and that their costs have fallen faster and more persistently than those of meat and milk. This is indeed what one would expect to find, bearing in mind that many crops have benefitted from the mechanization of field operations, from the multiplication of irrigation works, and from a supply of relatively cheap fertilizers. The livestock sector, however, has enjoyed no parallel advantages.

In these circumstances one would expect stock farmers to switch to crop production, but in few regions is this physically practicable, and in any case it is forbidden by the agrarian legislation, unless one sells his ranch and buys a crop farm. So profits for stock farmers have been squeezed. However, since for cattlemen the production cycle is long-term, most of them stay in business hoping that sooner or later the economic climate may become more favorable.

As to the future, it is accepted that price policy could be used as a stimulus for accelerating the growth of output. But in official circles it is argued that if prices were raised sufficiently, the consequence would be a brake on consumption, putting protein-rich foods beyond the reach of lower income consumers. The alternative of reducing production costs is alleged to be impracticable because feed crop prices have to be kept high in order to sustain the incomes of the hundreds of thousands of mini-farmers. We are thus in a vicious circle, and the dilemma is indeed unsolvable as long as we continue to examine it in the context of a closed economy.

Is it necessary to accept this constraint? Has it not clearly emerged from the model in this chapter and from the one in Chapter 2, that Mexico cannot hope to remain an economy characterized by self-sufficiency in food supplies? If this diagnosis be correct and if it be accepted that an import policy for agricultural products has to be formulated, then one of the strongest candidates for importing is the group of commodities needed for feeding livestock. Better to import the raw material than the finished product.

In a number of other countries governments elaborated arrangements under which they continued paying high prices to domestic producers of fodder cereals and imported the remainder of their requirements at the generally lower prices prevailing on the world market. The home supply and the imports were then sold, at a weighted average price, to the feed manufacturers, and as a result the livestock farmers could buy concentrates substantially cheaper than when the only source was domestic grain. They in turn could sell their end-products at lower prices, thus enabling consumers to increase their purchases.

Once the relationship between feed and end-product prices becomes economically attractive, mini-farmers can be encouraged by extensionists and helped with credit to enter the business, reduce their dependence on corn and beans and instead grow fodder crops for their own animals. By transforming feed into meat and milk the mini-farm can become a mini-factory. In this way a vicious circle could be transformed into a favorable spiral. Something along these lines urgently needs to be worked out for Mexico's farmers.

There are other obstacles which also will have to be removed before animal husbandry can prosper, notably in the fields of farm credit and of agrarian legislation, but these topics will be taken up in subsequent chapters.

The Institutional Framework

It is perhaps a platitude to say that in any society the institutional framework plays a major role in determining how individuals perform their tasks and what successes are achieved by the community as a whole. One framework may stimulate achievement, another may discourage it. In the farming sector one set of institutions may encourage producers to work hard to improve their land and their animals, their productivity and their output; another set may debilitate decision-making and lead to stagnation.

The management of livestock, and especially of cattle, in some ways resembles the management of an orchard. It requires several years for a herd of animals to become established and to attain full production. Decisions are taken by the producer on assumptions as to what the situation is likely to be in seven to ten years' time. If the prospects are favorable and conditions of stability can be expected to prevail, then a producer prepares plans for expansion and puts them into execution. If, on the other hand, uncertainty prevails, either nationally or for the individual farmer, the decision is likely to be to play safe, refrain from new investment, perhaps even to overexploit and deliberately run down the stock of productive capital. Unfortunately this latter is the present situation in Mexico.

What the farm sector needs is an institutional framework which encourages animal husbandry in general and livestock on mini-farms in particular. In this matter it will be convenient to consider separately three groups of producers: the large private farmers, the smaller private farmers, and lastly the ejidatarios and comuneros, because the needs of each group are somewhat distinct.

By large units we mean in the present context those with more than, say, 200 head of cattle or an equivalent number of smaller animals. Such units are to be found throughout the northern regions of Mexico and also to a limited extent in the south. The principal institutional deficiency for these ranchers is their insecurity of tenure. Even among those whose acres and cattle numbers fall clearly within the limits laid down by the law (and many surreptitiously exceed those limits), few have been able to obtain certificates of inalienability, because few possess written titles to their land and without titles such certificates are unobtainable. Anyone not having a certificate may at any moment find his land invaded by illegal squatters followed by the authorities who then expropriate all or a part of his property.

A certain number, less than 800 to be precise, with farms and herds in excess of the prescribed limits were given in the thirties and forties inalienability certificates valid for a period of 25 years. In 1965 it was decreed that no prolongations should be granted, and that upon the expiry of any particular concession, a team of agronomists and surveyors should be sent to arrange for the transfer to ejidos of any excess of land and/or animals. However, years have passed and few ranches have been subjected to this treatment; the remainder continue in a state of uncertainty, insecurity, and illegality.

Mention was made earlier of the legislation which originally prohibited ranchers from growing any crops but which was modified in 1971 to allow them to grow a limited quantity of fodder for their own use. It still is forbidden for one rancher to help another when the first has a feedgrain surplus and the second has been hit by drought. It is still forbidden for a crop farmer to add a herd of cows to his enterprise. And in any case the procedures involved in obtaining an inalienability certificate for "fodder-ranching" are so complicated that few farmers have been able to comply with all the requirements.

Of the small and medium private farmers in the livestock sector (those having less than 200 head of cattle or its equivalent), obviously very few of them would be in danger of breaking the law regarding the permitted number of acres, but they nevertheless have problems. Most of them suffer from lack of titles and hence from lack of inalienability certificates, and without these certificates they have no defense against arbitrary intervention. They are just as vulnerable to invasions as are the owners of larger farms, perhaps more so because the smaller farms often have richer pastures and better quality cattle. They too cannot grow forage crops without securing the special certificate. It is impossible for one man legally to operate two separate farms, one for livestock and the other for crops, and work the two as an integrated unit. It is also impossible, however small the scale of the enterprise, to obtain a certificate to practice commercial mixed farming, selling both crop and livestock products, although this would spread the risks, utilize more fully the family manpower, and cut down the production costs of meat and milk.

Let us examine finally the situation of the ejidatarios and comuneros, by far the largest class of farm operators. In the agrarian legislation there is persistent ambiguity as to whether the provisions in the articles govern the ejido as a unit or the ejidatario as an individual. Obviously those who drafted the legislation assumed that ejidatarios do not possess animals because nothing is said about any upper limit to the number an individual may own. In some of the large pasture ejidos in Chihuahua it would be perfectly feasible for a single ejidatario to own more than the permitted 500 head of cattle. The fact that few such cases occur is mainly due to the reluctance of the Credit Bank to help ejidatarios with livestock activities, which in turn is one of the main reasons why ejidatarios are on average poorer than private farmers having similar amounts of land.

Under the regime of President Echeverría the pendulum of policy swung the other way. Instead of ignoring the animal husbandry activities of ejidatarios and comuneros, the authorities began to offer generous inducements to ejidos to establish collective livestock units, and credits were granted for herds of 120 up to over 1000 or more cows or beef cattle. But the management of such units poses a number of technical and administrative problems. First, it is difficult to maintain disease control effectively when many animals are housed under a single roof, and this is especially so where the operators are largely uneducated. Secondly, the feeding costs are relatively high for a large herd concentrated in one place. Alfalfa and other bulky crops have to be transported long distances; no use is made of crop residues and kitchen waste which are the standbys of the family farmer. Thirdly, the care and management of the animals is in the charge of an employee of the collective who works his hours and then goes home regardless of whether a cow may be going to have difficulties dropping her calf during the night; whereas the individual operator has an economic incentive to tend his animals at all times.

The history of experiments in collectivization has been a long one in Mexico and has been no more encouraging in livestock than in crop production. Probably some will struggle on, their debts forgiven and their operations passing into the hands of a zootechnician appointed by the Credit Bank. But, as long as the countryside is overpopulated, the more rational policy is to concentrate aid on the small family producers.

In this chapter the recent developments in the livestock sector have been reviewed and the perspectives for the future evaluated. The neglect of the sector, reflected in the disproportionately small allocations of funds to livestock research and to the training of livestock extensionists, the latent hostility to livestock farmers mirrored in the insecurity of tenure for the larger ones and the absence of credit facilities for the smaller ones, combined with price policies which make animal feedstuffs too expensive—all these factors have created a situation which has impeded development and has caused the per capita consumption of meat and milk in Mexico to be one of the lowest in Latin America. This in turn has had detrimental effects on the nutritional status of large sections of the population. It is simply not in the national interest that these shortcomings be allowed to persist.

However, if the nation's nutrition is to be improved through a more adequate intake of animal protein and more liberal supplies of milk for children, the livestock industry has to be expanded. Population increase alone requires a doubling of supplies in 25 years, and if per capita consumption is to increase as well, then a still larger expansion is called for.

At the moment the country is poorly equipped for such a task. A first requisite is to establish two major research activities, one for animal breeding and one for pasture improvement. A second is to formulate a new policy for agricultural imports, recognizing the impossibility of self-sufficiency and facilitating the import of cheap feeds rather than expensive milk products and other prepared foods. The new feeds policy should be supported by starting up the domestic manufacture of petroleum protein. Next a thorough reconsideration of the current agrarian legislation is imperative. Its crude separation of crop from livestock farming no longer reflects the country's needs nor relevant technology, and should be replaced by texts which facilitate the practice of mixed farming. Above all, all farm operators, whatever their scale of business, need to be assured greater security of tenure without which there will be no investment in improvements. How this could be done will be considered in a later chapter.

The livestock sector can make perhaps the most important single contribution to improving the diets of the Mexican people and to bettering the incomes of Mexico's farmers. If this is to be accomplished, there will indeed have to be elaborated programs of the necessary magnitude; but more than anything else there is required a change of approach, in the sense of creating more positive attitudes toward the whole sector on the part of professionals, political leaders, and the general public.

6. Technology and Productivity

In the three preceding chapters we analyzed the basic physical resources at Mexico's disposal for agricultural production—the amount of cultivable land, the irrigation facilities, and the livestock. It became apparent that in none of these is the supply increasing, or capable of increasing, as fast as the national demand for food and agricultural raw materials. Therefore, it is important to examine what contribution technology could make toward resolving this very real dilemma, trying to answer the questions: to what extent has the technical efficiency of farming been improving, and what are the prospects for further gains?

In the farm sector the process of applying technology to production involves a number of stages. First, there is agricultural research, then the transmission of the findings of research to farmers through an extension service and other media of communication. The impact of this propaganda shows up in the quantities of inputs used by farmers, and finally the managerial skill in combining inputs effectively becomes measurable in terms of the productivity of land, of animals, and of manpower. Each of these topics will now be examined in turn.

AGRICULTURAL RESEARCH

Agricultural research possesses several features which distinguish it from research in other branches of science. One is that only a minor part is basic research while the major part is applied. For example, the study of genetics is basic in the sense that its principles are universally valid, but the application of genetic science to producing, say, a new variety of corn will be successful only to the extent that it has been adapted to a particular zone's climate, soils, pests, diseases, and other local factors. The findings of basic research can be taken over from other countries, but applied research is location-specific; it has to be undertaken where the results are wanted.

For instance, CIMMYT (Centro de Investigaciones parà el Mejoramiento de Maíz y Trigo), the Rockefeller Center for corn and wheat research located outside Mexico City, screens the germ plasm in its world collection of wheats, undertakes a breeding program of two cycles a year (profiting from the contrasted climatic zones within the country), and then

[116]

distributes selected plant material for trials in all the major climatic regions of the world, where they are exposed to various types of soil and moisture conditions, day-length and temperatures, diseases, and insects. One result of this worldwide testing is the identification of materials with wide adaptability and yield stability. A similar distribution for testing has to occur within Mexico, because of the diversity of its soils and climates, especially for crops such as corn which are less adaptive than wheat.

Applied agricultural research is long-term in character. It takes several years to breed and establish a new variety of a crop; it may take more than a decade to produce a new breed of cattle adapted to a difficult environment. Another awkward feature is that rather little agricultural research can be wished onto private industry as is the case in, for instance, pharmaceuticals; most of it has to be government sponsored. Yet it is hard, especially in a relatively poor country, to sell agricultural research programs to political decision-makers, for it means asking them to sow seeds whose fruits will accrue only to a different government some years later.

Indeed, governments are often seduced by the argument that the most pressing problem is not additional research but rather the yawning gap between already available knowledge and the average technical competence of farmers. Hence they may give priority to expanding their extension services. This is a specious argument, because in fact particular research programs have to be initiated several years before the nation requires their results. Even in Mexico, where successive governments have shown a sense of responsibility in these matters, the funds allocated to INIA, the chief national research institute, have barely kept pace with the inflation in its operating costs.

Furthermore, the personnel problems may be as acute as the financial ones, because of the constant brain drain out of agricultural occupations. In the agricultural universities the courses inevitably have an academic slant, so that the brighter students expect to take masters' degrees and frequently doctorates. Having achieved their titles, many feel too grand for practical work in the laboratory or the field, still less are they willing to serve in the provinces. Their aim in life is to become bureaucrats, preferably in some federal agency where both the prestige and the perquisites are more considerable.

In Mexico, as elsewhere, the research directors are reluctant to admit the practical limitations in respect to financial and human resources. Hence they are inclined to pay insufficient attention to resource allocation. It is easy to argue that everything is important—all crops, all animals, all pests and diseases, and so on. There is no topic which is not of interest to at least some farmers in some corner of the country. For example, at INIA's regional research stations the programs embrace such a wide range of subjects that only one agronomist can be assigned to each group of crops, and thus he merely scratches the surface of his problems. If the programs were fewer and more selective, their productivity would be enhanced.

Program selection involves taking a long-term view of a country's agricultural needs. This is a complex task, and forward guesses may be wrong. Responsibility tends to be diffused among scientists, administrators, and politicians. Hence many countries drift along without making any choices, and consequently are saddled with programs suited more to the needs of thirty years ago than to those of the present.

Something akin to this has occurred in Mexico. Thirty years ago when the Rockefeller plant-breeding program was initiated, it was oriented chiefly toward wheat, because of the nation's then dependence on imported supplies. The wheat program achieved spectacular successes, not only in making Mexico self-sufficient in wheat but in exporting wheat varieties to other third world countries, thus starting what came to be called "the green revolution." Meantime the scenario was changing. Population was expanding, incomes were rising, and patterns of consumer demand were modified. Also the quantity of cultivable land and of irrigation were nearing their limits. Although wheat production expanded more than sixfold between the mid-forties and the mid-seventies, it was all absorbed in consumption. In corn and oilseeds Mexico became an importer; in sugar the demand could be met only by switching from exports to imports. Livestock production became more costly, putting milk and meat out of reach of the lower income-groups, because livestock research had been neglected. These were developments which the plant geneticists could not be expected to monitor. The decisions lay elsewhere and were not taken early enough to meet the new challenges.

Certain critics have accordingly attacked the agricultural research authorities, alleging that the programs mainly serve a privileged class of farmers (those in the irrigated districts) and calling for more research on corn for cultivation on non-irrigated land in areas of low rainfall where most of the poorest farmers are located. Actually, research on corn varieties has been included in the Rockefeller program since its inception, but striking results are inherently more difficult to obtain. Hybridization presents its distinctive problems—only one cycle can be grown per year, and there are many mini-ecologies to be adapted to, each with distinctive peculiarities. Already available are dwarf varieties which produce high yields under irrigation. It is reasonable to expect that the great efforts recently focused on varieties for low rainfall regions will be producing satisfactory results in ten years' time; and these blessings will arrive just when what the nation needs will be not corn but other commodities.

The unfortunate fact is that decision-makers are preoccupied with the present and fail to prepare, or prepare incorrectly, for the future, putting research scientists in a false position—like generals who are always fighting the battles of the last war but one. For instance, soybeans and safflower were introduced into Mexico by individual farmers in the early sixties; yet not till many years later did the regional research stations begin to include these crops in their programs. In the seventies some farmers in the northwest began growing table grapes and certain types of nuts, but applied research on the local problems of these new crops is unlikely to be inititated before the mid-eighties. Instead of trailing along behind the farmers, the research stations should be taking the lead in trying out novelties and introducing the successful ones to the farming community.

It is not too difficult to delineate in general terms what the consumers will be demanding say ten years hence, as we did in Chapter 2. Although the rapid expansion of population will occasion an all-round increase in requirements, demand will grow fastest for the more expensive foods, such as livestock products and certain fruits and vegetables, while there will probably be export opportunities for coffee, cacao, and some tropical fruits. Since Mexico will be becoming a larger importer of several items, it will pay her to discontinue importing large amounts of milk products, buy more animal feed from abroad, and expand her own milk production.

What does a scenario of this kind have to suggest in regard to research needs? First of all, top priority and a major allocation of funds should be assigned to *animal breeding programs*, both because they have been relatively neglected and because of the long time-lag before they give results. These will be oriented toward beef cattle, dairy cattle, and dual purpose animals, each and all adapted to tropical and subtropical areas; projects of sheep and goat improvement for the marginal Altiplano grasslands, and a program to reduce the number and improve the quality of breeds of hogs.

Breeding programs require support from research in feed supply improvement. One such improvement program should focus on experiments and trials with exogenous grasses from other countries having similar climates and soils, as proposed in the previous chapter, partly to restore and improve the rundown grasslands and partly for inclusion in crop rotations in the tropical forest zones of the south. Another should be devoted to oats, feed barley, peas, feed corn, and, to a lesser extent, the oilseed crops. In this group of commodities little systematic work has been undertaken, and per-acre yields are discouragingly low.

The other group of products requiring research are those destined for export markets. It is extraordinary that after more than a century of coffee and cacao exporting, so little research has been done, so that only a small proportion of either crop reaches export standard. Research initiatives are also overdue on the exportable fruits and nuts with attention to their endurance of long distance transportation. These omissions occur partly because reponsibility is ambiguously divided between the Department of Agriculture's stations and those of the specialized bodies, such as the Mexican Coffee Institute. Another need is for more rigorous inspection and control of the commercial enterprises which supply planting material and which are all too lax in respect to the quality of what they sell.

To make room for these programs within the budget of plant-breeding research as a whole, the programs ongoing for other crops might have to be diminished, for example those on cereals for human consumption. This may be regrettable, but such pruning is inherent in any attempt to establish and implement priorities. There are other products, henequen being a notable example, which desperately need research because yield and quality are so low, but which from a market viewpoint have such a dismal future that an allocation of funds to this task would be unjustifiable.

Plant-breeding research requires the support of other programs directed toward such topics as water use, fertilizer application, pest and disease control, machinery and implements, farm management, and so on. Thus, although water is so scarce in Mexico, even in most of the irrigation districts, one does not encounter programs designed to test methods of economizing water use. Much could be learned from other parts of the world, and tried out in Mexico, regarding irrigation techniques, the timing of applications, and the conditions under which supplementary irrigation would suffice.

In the matter of fertilizers, data on crop responses to differing mixtures and quantities are quite limited and mostly date from many years back. Meanwhile the extensionists continue to recommend a few standard formulae, regardless of the peculiarities of crop or region. They also continue to maintain the strange prejudice that Mexican soils do not require potash, though certain recent experiments have suggested that potash, although indeed present in most of the soils, is of a kind which plants do not assimilate. Be this as it may, what is needed is the establishment of several hundreds of fertilizer trials distributed through all the main crop-growing regions.

The use of farm machinery in Mexico also requires reevaluation. Farmers everywhere like to buy as much power as they can afford, irrespective of whether they really need it, and Mexican farmers in their purchases of tractors, for example, have imitated the North American predeliction for massive power. But American farms are large while most Mexican ones are small and do not need the 80 to 120 horsepower machines which prevail. Most farmers could operate just as efficiently and more cheaply with 20-30 hp machines, or even with garden tractors. Simultaneously, much useful work could be done to modernize the design of hand tools: the machete is not quite the universal answer that Mexican tradition pretends.

Finally, a wholly neglected field is that of farm management. One step would be to devise practical and profitable combinations of crop production and livestock suited to small farms, with a view to overcoming the traditional divorce between these two activities and enabling farmers to maximize their gross output from small acreages. Another suggestion would be to devise and introduce a simplified form of farm bookkeeping, so that farmers would begin to think more rationally about the financial relationships between inputs and output. At the end of the seventies the Department of Agriculture began collecting information via sample surveys on the production costs of a few selected crops, but with the purpose of determining adjustments in the levels of guaranteed prices. And since these surveys did not relate differing levels of input expenditure to different yield levels, much more would be needed as a guide for advice in farm management.

The principal institution charged with the conduct of agricultural research is the Instituto Nacional de Investigaciones Agricolas (INIA), founded in 1960 and possessing eight regional research centers, 40 field stations, and 11 substations. Altogether it employs nearly 500 investigators. Over the years there has been continual controversy about centralization or decentralization of its activities. The centralists emphasize the need for tight control of operations and quality standards and thus prefer concentration at Chapingo adjacent to the agricultural university. Decentralists insist on the location-specific character of most agricultural research.

The argument, of course, should concern not *whether* to decentralize but *what* to decentralize. Obviously, any basic list of priorities must be centrally determined, likewise the main features of the program in each subject field, as also ex-ante evaluations of alternative ways of pursuing the program objectives. But field operations need to be carried out in as many different locations as financial resources permit. A single state of the Republic may possess several quite distinct ecological zones, such that the results of field experiments cannot unquestioningly be applied from one zone to another. Guerrero, Oaxaca, and Veracruz, to name only three states, each have a number of mini-regions.

Decentralized research also enjoys the advantage of direct contact with farm people. In a country of traditional practices, such as Mexico, the local people have learned over the centuries what the local adversities are and how in some measure to overcome them. The investigator sent from headquarters is unaware of these matters and sometimes is too proud to learn about them.

The traditional farmer is averse to risk taking. He can be persuaded to depart from his customary ways only if he sees (on some piece of land) the almost certain prospect of substantial gain—a mere 15-20 percent yield improvement will not convince him. In one instance in the state of Puebla field trials showed a large yield gain for corn and beans grown in combination resulting from a certain fertilizer formula. However, the farmers rejected the advice because the application would have cost 20 dollars per acre in fertilizer, and if the rains did not arrive at the right moment the outlay would have been in vain. They said: we are prepared to risk not more than five dollars per acre, so give us a fertilizer mix within that limit.

Research administrators tend to forget that research staff should not consist solely of agronomists, plant pathologists, and all the other specialists but should also include participating farmers. For one thing, the results of trials on farmers' plots are frequently more realistic than those obtained at field stations where conditions are too nearly optimum. For another, as a consequence, the farmers who participate become extension workers almost involuntarily; they are the most convincing sellers of what has succeeded in their own experience. (This was the experience, for example, in a corn improvement program operated in the state of México.)

That more research is not decentralized is due largely to shortage of funds; and a significant part of this shortage stems from the wasteful overlapping of responsibilities between official agencies. In Mexico at least thirty distinct agencies conduct agricultural research, and among these a single agency may administer a large number of programs. Mexico is not unique in possessing this duplication and dispersion of effort, but it is more deplorable where there is such scarcity of trained personnel; as a result one tends to find low levels of technical performance in the less powerful institutes which outweighs any advantage that might be gained through interagency competition. What is needed is an impartial review commission to examine the whole picture, eliminate redundant agencies, and redistribute their tasks among at most three or four.

AGRICULTURAL EXTENSION

Agricultural advisory services to farmers began to be formally organized in Mexico as long ago as 1911, though at first they were provided by a tiny group of agronomists. Gradually the service grew; in 1948 it acquired the title of "agricultural extension," and in 1971 it attained the rank of a department within the Department of Agriculture. A few of the individual states of the Republic have extension services of their own, but most confine their support to providing facilities for the federal government's program. As in the case of research, other agencies too have become involved in this type of service, so that there exists the same overlapping and confusion.

Among these others, by far the most important is the Rural Credit Bank, formed in 1976 by a fusion of the three former agricultural credit banks. This bank, government owned and operated, is the principal provider of farm credit, and inasmuch as most of this is supervised crop credit, the bank employs a large staff of extensionists as supervisors. Their tasks include inspecting the farm before an application for credit is approved, helping the farmer obtain seeds and fertilizer (often the bank buys these for him), arranging the hire of a tractor, seeing that he performs his weeding and spraying tasks and advising him on the disposal of his harvest (sometimes the bank sells it for him). In this way the bank's extensionists become more directly involved in farm operations than those of the Department of Agriculture.

The extreme case is that of the collective ejido. When a large ejido, say of 50,000 acres, becomes collectivized, it needs substantial finance for farm buildings, roads, machinery, etc. and turns to the Rural Credit Bank. Because the scale of operations is large, the bank appoints the necessary senior staff: a manager, an accountant, a resident agronomist, perhaps a livestock man. These are responsible to the members of the collective through its general assembly and to the bank on financial matters.

The extensionists of the Department of Agriculture also undertake regulatory duties. For instance, they have to ensure compliance with the requirement to report notifiable pests and diseases and to destroy condemned crops or animals. Likewise, they distribute and enforce the local acreage quotas for crops under statutory limitation. Inevitably this work involves them in disputes and frequently impairs their capacity for influencing farmers to adopt improved practices. It is desirable that the two tasks be separated and the regulatory duties undertaken by a different staff of inspectors.

The extension workers in Mexico face formidable obstacles. Illiteracy in rural areas averages over 50 percent and in the more isolated districts between 80 and 90 percent. In many such districts the farm people have no Spanish and speak only their local language dating from pre-Columbian times. The more assiduous extensionists try to learn the language prevailing in their area (though they get no bonus for doing so), but they may suddenly be transferred to another part of the country. In certain districts the hostility toward employees of government agencies is so great that the extensionists keep out rather than be subjected to violence.

Resistance to modernization comes in Mexico, as elsewhere, mainly from the older generation, whose attitude to pest and disease control, for instance, is: "God gave the plague, God will take it away." Many feel it would be disloyal to their ancestors to depart from inherited customs. When a son is already managing the farm, he may tell the extensionist that he accepts the new techniques but does not wish to introduce them till his father passes away. Meanwhile, however, the extension workers might be doing more to mobilize the interest of the younger generation. One finds very few 4-H clubs or similar activities, few local competitions and shows.

Nonetheless, most of the extension staff are devoted to their tasks, working long hours, including Saturdays and Sundays when the farmers are most accessible; and the same is true of the limited group of girls teaching home economics. In remote regions many villages can be reached only by a six-hour mule journey. It does not promote good feeling between agency personnel if, when the Agricultural Department's extensionist arrives at such a village, he finds that the extensionist of the Credit Bank was there the day before.

The extension staff is poorly paid, and receives quite inadequate travel expenses. More often than not, their salaries (and reimbursements) are several months in arrears. There are few opportunities for promotion, and no facilities are provided for enabling the brighter assistants to proceed to higher agricultural education. The senior extensionists are almost all specialists in one or another discipline; yet what the thousands of illiterate small farmers need is advice from generalists who would call in some specialist only for intractable problems.

Extensionists are instructed to serve ejidatarios and only visit private farmers if they have time to spare, which they never do. Thus they contribute to emphasizing the social cleavage between the two tenure groups, already a blight in the rural areas. Moreover, many private farmers are just as ignorant as ejidatarios and operate on an equally small scale. Surely the criterion of eligibility for technical assistance should be need rather than form of tenure.

Extensionists tend to receive too many directives from above and not enough advice from below. Their first task after all is to seek to understand the farmer's mentality and the reasons motivating his resistance to change. Especially in the marginal subsistence zones a socio-psychological approach is essential.

There are strong arguments for going further and incorporating some of the better farmers formally into the extension system. They can communicate more successfully with others of their village whereas the extensionist will always remain a "foreigner." Recruiting two or three brighter and younger men in each village and employing them as assistants on a part-time basis would in effect create a two-tier system, such as has proved useful in other third world countries. Like the barefoot doctors in China, their role should be that of missionaries; their farms would become demonstration plots which their neighbors would observe and gradually copy. Something of this kind seems worth trying, at first in selected districts, if the inertia and ineffectiveness of the extension service in its present form are to be overcome.

The effectiveness of research and its diffusion through an extension service should be reflected in the progress and modernization of a country's farm sector. Therefore, as indicated at the beginning of this chapter, our next step is to examine the extent to which Mexican farmers have been adopting modern practices. While there are several yardsticks which could be used for this purpose, we have singled out two, the use of fertilizers and the employment of farm machinery, as sufficiently symptomatic of the modernization process.

FERTILIZERS

Over the past thirty years in Mexico the rate of increase in fertilizer consumption has been one of the fastest in the world, although it started from a very low level. In fact, until the Second World War it was only Western Europe which had reached high levels of per-acre consumption. In other continents the spectacular expansion began only in the 1950s, partly because in relative terms the prices of fertilizers were falling and partly because in a number of countries the governments were subsidizing consumption.

TABLE 6.1

FERTILIZER CONSUMPTION IN SELECTED COUNTRIES:
POUNDS OF NUTRIENTS $(N + P_2O_5 + K_2O)$ PER ARABLE ACRE

	Pounds per Acre		Increase (1948–52 =100)
	1948–52	1977–78	
Argentina	0.45	2.0	436
Venezuela	0.86	40.2	4,678
Mexico	1.28	41.0	3,205
Colombia	4.75	45.7	961
Brazil	2.54	69.0	2,718
U.S.A.	20.2	88.8	439
France	44.8	247.6	553
Germany (Fed. Rep.)	147.6	376.4	255
Belgium	239.6	475.3	198
Netherlands	347.5	657.7	189

SOURCES: *1948–52:* Arable area from FAO Production Yearbook, 1953; fertilizers from FAO Production Yearbook, 1963.
1977–78: Arable area and fertilizer consumption from FAO Fertilizer Yearbook, 1978.

As Table 6.1 shows, even the European countries have registered increases from their already high levels, but the U.S.A. has quadrupled her consumption while several Latin American countries have increased theirs tenfold or more, and Mexico more than thirtyfold. It should not be expected that fertilizer use in Latin American countries will ever rise to the levels which have proved profitable in Western Europe—the ecological differences and the man/land ratio contrasts are too great—but there undoubtedly remains a great potential for further expansion.

Naturally one finds significant differences in the use of fertilizers in the various regions of the Mexican Republic. The only information available by states is collected in the decennial farm censuses and is by expenditure rather than by quantities. These indicate that expenditure per acre of crop land in Sonora was seventy times that in Quintana Roo. Consumption tends to be high in states which have irrigation or ample rainfall, which are oriented toward commercial rather than subsistence crops, and in which one finds a high proportion of innovating farmers.

Over the years these influences have been spreading, so that narrower differences exist than in the past. Also the rapidly growing demand for food in the major metropoli has had an effect: the states adjacent to Mexico City, to Guadalajara and Monterrey have become heavier fertilizer users, particularly in order to provide truck crops to the cities. With the official encouragement of industrial decentralization, more cities will achieve economic take-off, and cultivation in the farm areas around them will become more intensive.

The situation can also be considered by main categories of farmers. Thus at the 1970 census average fertilizer expenditure per acre was 110 pesos (9 dollars) among private farmers with farms above 12.5 acres, 70 pesos on those below 12.5 acres, and only 51 pesos on average among ejidatarios and comuneros. This might be explained if these latter two groups possessed less than their proportionate share of the irrigated land, but such is not the case. One valid reason for their lesser consumption is that these ejidatarios and comuneros are not so deeply involved with high-value crops that in general require much fertilization, but probably more important is their lower level of education and agricultural know-how.

Strangely enough, in the irrigation districts we find little difference between ejidatarios and private farmers, as the figures in Table 6.2 indicate. The ejidatarios fertilized a larger proportion in the northern region and the private farmers a larger one in the south, but otherwise the differences were minimal. When the performance is broken down according to crop, the

TABLE 6.2

**PROPORTION OF CROP AREA RECEIVING FERTILIZER
IN IRRIGATION DISTRICTS:
1976–77**

(percentage of total crop area)

Region	Ejido Farmers	Private Farmers
Northwest	87.8	90.5
North	85.4	84.7
Northeast	28.7	57.0
Center	76.7	69.9
South	41.2	72.7
All districts	76.5	78.9

SOURCE: Department of Agriculture, Bureau of Agricultural Economics, Agricultural Statistics of Irrigation Districts (annually). Regional groupings of states for irrigation districts differ slightly from those used in this book.

ejidatarios fertilized slightly more of their cotton than did the private farmers, and the latter slightly more of their corn and wheat; but again the differences were insignificant.

One reservation has to be recorded concerning these conclusions. In the irrigation districts, as noted in Chapter 4, the ejidatarios illegally rent a large amount of land to private farmers. The operations on these rented lands, although carried out by private farmers, are attributed to the original occupiers, whether this be fertilizer use, machinery, yields, or output. As a result, in these cases and for these reasons the technical efficiency of the ejidatarios tends to be exaggerated.

The same special survey of fertilizer use in irrigation districts indicated that at that time (1972) the overall average consumption per crop acre in all the districts was 290 pounds (product weight) but reached 422 pounds in the districts of the north and central regions. This means that in those zones the consumption level had attained already the French national average. Although Mexico's irrigated areas account for only a little over 20 percent of the total crop area, they take more than half the total quantity of fertilizer used.

There are practical reasons for the contrasts in fertilizer use as between the irrigated and the other lands. One is the uncertainty of the rainfall. Since a sufficiency of moisture must be present in the soil for fertilizers to produce their effects, if a farmer applies fertilizer and then no rains come, he has largely wasted his money; and as we have seen, the more traditional Mexican farmers are averse to taking risks of this kind. The other practical reason is the absence of knowledge of crop responses in rain-fed areas for lack of systematic fertilizer trials.

It is sometimes argued that the prices of fertilizers are too high for the smaller-scale farmers to be able to afford them. This contention is unsupported by the evidence. Investigations have shown that expenditure on fertilizer is around 10 to 15 percent (according to crop) of total production costs, and almost invariably a fertilizer application would augment the crop's yield by more than this percentage, so that the outlay would constitute a good investment except in areas of very uncertain rainfall. Moreover, since price data began to be published in 1960, fertilizer prices in Mexico have risen less than the general price index; that is, they have fallen in real terms, partly due to subsidies. Nor are they expensive compared with the prices of the same products in other countries. Up to now the main obstacle to their wider use has been the technical competence of the smaller farmers.

This leads us to speculate on the outlook in the Mexican fertilizer market during the 1980s. During the late seventies consumption was growing at around 13 percent per year, somewhat faster for phosphates than for nitrogenous fertilizers. In the irrigation districts any very rapid expansion is unlikely, since many farmers are already employing optimum quantities. In the arid and semiarid zones of uncertain rainfall, very heavy applications will never be a commercial proposition. The main potential for market expansion is likely to be located in the non-irrigated areas of adequate rainfall, for instance all down the Gulf coast and in some parts of Jalisco and Michoacán. For many crops in these regions it would probably pay to triple or quadruple fertilizer applications, though we cannot be sure until more experimental data are forthcoming.

Bearing in mind that in the coming years the Mexican food balance will become increasingly dependent upon imports, it seems reasonable to suppose that the government will pursue a policy of maintaining a favorable cost/price relationship for farm products, and therefore the inducement to modernize and intensify cultivation practices will continue to be strong. This suggests that the market for fertilizers may continue to expand at 13 to 15 percent per year with perhaps some deceleration toward the end of the decade. The Mexican fertilizer industry is quite capable of sustaining such a rate of growth in respect to nitrogenous and phosphatic fertilizers; only potash has to be imported and, as already noted, it is not widely consumed.

FARM MACHINERY

From the beginning of the twentieth century and for several decades, machines and horses worked side by side on farms, partly because the machines had not yet been sufficiently adapted to all agricultural operations and partly because farmers loved their horses more than they could ever love a tractor. However, since the 1930s in North America and since the fifties in Europe, the machine has achieved its final victory, and work animals have been almost entirely eliminated.

In the third world countries the same process is taking place, though it is still at an earlier stage. In some the number of work animals is already declining considerably (for instance, in Thailand, India, Argentina and Colombia), while in others the number is still rising. In all, the tractor population is growing rapidly, though it is still small in absolute numbers. Mexico is one of the countries in which horse and mule numbers have continued to rise.

Apparently during the past thirty years the amount of traction power on Mexican farms in relation to the area of crop land has increased significantly, taking animal and machine traction together. Moreover, the figures underestimate the true increase, since there has occurred a substitution of horses and/or mules for oxen, a process which increases the work efficiency per animal. Only in the poorest and remote districts do oxen still perform most of the field work (Oaxaca and Chiapas for instance).

The question arises whether the amount of power available on farms is adequate or excessive. Mexico, Colombia, and Peru are the most heavily stocked with horses and mules on the American continent, with more than 100 head per thousand acres (Table 6.3). In tractors no Latin American country possesses more than 3 per thousand acres, the U.S.A. has 9 but the Federal Republic of Germany has 73 and Japan 77. Combining the two components, with a rough and ready coefficient of 20 horses equalling 1 tractor, Mexico has about 11 "tractor units" per thousand acres of crop land, the U.S.A. 10 and the Federal Republic 74.

The figures in themselves mean little until they are related to the local situations. Obviously a larger amount of traction power is needed on heavy soils than on light ones, on steep hillsides than on flat land, on small fields

TABLE 6.3

HORSES, MULES, AND TRACTORS PER THOUSAND
ARABLE ACRES: SELECTED COUNTRIES, 1977 AND 1978

Country	Horses & Mules (1978)	Country	Tractors (1977)
Mexico	169	Japan	77
Colombia	157	Germany (Fed. Rep.)	73
Peru	102	France	30
Brazil	76	United Kingdom	29
Venezuela	41	U.S.A.	9
Argentina	33	Canada	6
U.S.A.	21	Australia	3
Germany (Fed. Rep.)	19	Brazil	2.8
Indonesia	15	Indonesia	2.7
France	9	Mexico	2.6
United Kingdom	8	Venezuela	2.3
Australia	4	Argentina	2.2
Canada	3	Colombia	1.8
India	2.5	Peru	1.5
Japan	2	India	0.6

SOURCE: FAO Production Yearbook, 1978.

(with constant turning) than on large. Thus a country in which heavy soils, steep slopes, and small farms predominate will require more traction power than one having the opposite characteristics. Much of Mexico suffers from these unfavorable factors, indicating that much power would be justified; yet she possesses less than the U.S.A. (per thousand acres).

Here, however, another factor enters. It is a general experience that when farmers become prosperous they equip themselves with more traction power than they really need. In the old days it gave a man prestige to own several horses even if he needed only two for his farm work. Today he likes to have several tractors, the outstanding examples being among the Germans and the Japanese. Of course a tractor can break down in the middle of harvest, so it is preferable to be over rather than under "tractorized"; and for a commercial farmer the additional amortization costs are immaterial.

This tendency to overstock with traction does not harm the nation's food supply so long as the main component is the tractor, but when it is mainly animals, as in third world countries, the pressure may be considerable on the available pastures and on the supply of other feeds. In Chapter Five we noted how in Mexico the proportion of non-productive animals has been increasing in some of the poorer regions and is particularly high among the ejidatarios. This reduces their scope for expanding dairying and sheep and goat production.

In Mexico at the last count (1970) just over half the crop area was still being cultivated by animal power; even on the larger private farms the proportion was 25 percent. Much lower percentages were registered in the technically advanced northwest region where even the ejidatarios were more than 70 percent mechanized. On the other hand, in the central states the ejidatarios used animal power on 90 percent of their crop land. Interestingly, the small private farms of less than 12.5 acres in size had little traction power of their own; they primarily hired either mules or a tractor for their cultivations and harvesting.

In the northwest the larger farmers in the irrigation districts reckon to have about one tractor for each 150 acres of crop, though many own more than this for the reasons mentioned. In the poorer regions of rain-fed farming, a farmer with 8 to 10 acres will typically have a pair of mules or oxen. In the irrigation districts, according to a survey conducted in 1972, some 61 percent of the private farmers' crop land cultivation was mechanized and 49 percent of that of the ejidatarios; in the rest of the country the corresponding figures were 27 and 11 percent respectively. But irrigation districts are composed of flat rectangular plots, whereas the other areas tend to have much more awkward topography.

In looking forward from these various experiences toward what may be expected to occur in the coming years, we have to note a certain ambivalence inherent in the mechanization process itself. We have to ask how far it is desirable in a country such as Mexico, with so much real and disguised unemployment in the farm sector, to introduce more and more machinery inevitably reducing the volume of employment on the land. This dilemma has several distinct aspects which here can only be summarized.

Several advantages seem obvious. (1) Farm machinery has the capacity to increase labor productivity and hence increase the incomes of farmers and their employees. (2) Tractors, working more rapidly than animals, can take advantage of favorable weather conditions for sowing or harvesting, and consequently increase crop yields. (3) By eliminating work animals, land is liberated for growing food for human consumption. (4) Mechanized cultivation reduces the quantity of physical energy expended by the farm worker. (5) Driving a large machine gives its operator a fine feeling of power.

But mechanization has its disadvantages. (1) It involves more cash outlays than animal power: fuel, repairs, and in due course replacement. A horse or mule lives off small areas of grass and reproduces at no cost. (2) The animal is more versatile and can be used for various transport jobs. (3) The animal is a more reliable servant, whereas the tractor may break down at a crucial moment and repair facilities may be far distant. (4) As already emphasized, mechanization aggravates the already excessive unemployment.

Mexican governments over the years have pursued policies which encourage mechanization. Through the Rural Credit Bank they provide subsidized credit for the purchase of tractors and other machinery, they subsidize the price of the fuel and electricity bought by farmers, and through the regime of guaranteed prices they maintain farmers' incomes at levels which permit a steadily increasing proportion to mechanize. It could be argued that such policies are not fully consistent with parallel policies designed to promote employment.

Not that mechanization should be formally forbidden, even if that were feasible. Farmers will want to mechanize to the extent that they can afford it, and they cannot be kept poor just to discourage such a trend. However, it would seem desirable to reconsider the subsidies with the aim of gradually phasing them out.

Against this background we may briefly comment on the likely changes in the coming years. For example, it is certain that the use of oxen will continue to decline in the northern regions and will begin its decline in the southern ones. The horse and mule populations may be expected to reach their peaks before 1990. By that date the tractor population should exceed half a million. Simultaneously the use of milking machines and electric motors of other kinds will become widespread in the zones oriented to commercial farming. By contrast in the remote and extremely marginal areas it is more likely that farms will be abandoned than that they will ever become mechanized.

FARMERS' EXPENDITURES

Further light can be shed on the modernization process by a brief analysis of the size and composition of farmers' expenditures. At the subsistence end of the farm spectrum such expenditures are almost nonexistent, while a commercial farmer specializing in high value crops may have a very considerable cash flow.

Data from the last census reveal the huge contrasts in this respect between the more advanced and the more backward parts of the country. Thus, in Sonora and Baja California average total expenditure per acre of crop land was ten times that of Zacatecas and Guerrero. In expenditure on fertilizers and pesticides the difference was more than twentyfold. It is probable that at the beginning of the 1980s these gaps have narrowed very little.

Very distinct levels and patterns of expenditure emerge from the different groups of farmers. Estimates for the period 1979–80 indicate that the private farmers operating more than 12.5 acres had expenditures averaging 80 dollars per acre of crop land, those with less than 12.5 acres 220 dollars, and ejidatarios and comuneros 26 dollars. The high level of expenditure among the small private farmers is largely explained by the existence of large specialized poultry units in this group, and indeed these farmers devoted on average more than half their spending to purchases of feed.

What is striking and surprising is the low level of expenditure by ejidatarios, most of whom have farms smaller than 10 acres. This means that few of them have taken up intensive poultry or hog raising or have specialized in truck crops, policies which would have enabled at least some of them to obtain acceptable incomes from very small acreages. Instead the great majority continue to concentrate on the cultivation of corn and beans which, after satisfying the family needs, produce a surplus of low cash value.

Also surprising is the pattern of ejidatarios' expenditure. Over the country as a whole more than a third of their expenditure goes to the hiring of labor, while in the poorest states (e.g., Oaxaca and Guerrero) the proportion rises to 60 percent. It is frequently asserted in the agrarian literature that the ejidatarios have so little land that they are underemployed for most of the year; and one would have thought that on a 10-acre farm the operator and his family could cope adequately even at harvest time. Yet many of them prefer to hire workers with themselves sitting back supervising. The author visited one such farm where the cotton had gone unpicked because the able-bodied operator could not find anyone to come and do it for him. Where almost all the available cash is devoted to hiring, there is none left to buy fertilizers.

LAND PRODUCTIVITY

The modernization of farming can be traced from research, through extension, through the changes in farm practices as instanced by mechanization and wider use of agricultural chemicals, and finally becomes expressed in higher crop yields and other improvements in the productivity of land. Over a period of years we expect to observe an upward trend in physical crop yields, though this may be moderate or rapid depending on the crop and the locality.

Unfortunately, in Mexico the information available on this topic is none too reliable. The census data cannot be used because these refer to a single year once a decade. The Department of Agriculture does publish annual statistics dating back to 1925 for the area, yield, and production of a considerable number of crops, but problems arise in interpreting this information. In many products the time series is vitiated by revisions in the method of crop reporting. A particular crop may be showing an upward yield trend of, say, two percent per year, when suddenly the reported yield rises by 50 percent or more and thenceforward resumes its two percent trend. Adjustments of this nature have been made from time to time in the yields of such crops as wheat, cotton, potatoes, tomatoes, and strawberries, to mention only a few.

Since almost all the revisions have been in an upward direction, this has produced an exaggerated rate of improvement over the long term. For many important crops the annual rate of yield increase has averaged more than three percent over a period of fifty years—a result difficult to believe since in other countries which have adopted technology more rapidly and more completely a growth rate of two percent compound is considered a meritorious performance.

Nevertheless, utilizing the figures with caution, it may be legitimate to draw certain conclusions from the accompanying graph (Fig. 6.1). A striking contrast is immediately apparent between certain cereals, vegetable crops, and cotton on the one hand and most fruits and plantation crops on the other. The former have all improved significantly, even after discounting the distortions in the trends, whereas the latter crops have barely improved at all. Indeed there are some, such as bananas, henequen, mangoes, and apples, whose yields in the late seventies were lower than they were fifty years earlier.

A partial explanation of the deterioration may be found in the rapid expansion of the area devoted to these crops, because it is the common experience that any particular crop is first grown on the soils and in the climate most suited to it; then when its cultivation is expanded in response to

FIGURE 6.1

TRENDS IN PER-ACRE YIELDS OF SELECTED CROPS: FIVE-YEAR AVERAGES, 1925–29 to 1975–79 (1925–29 = 100)

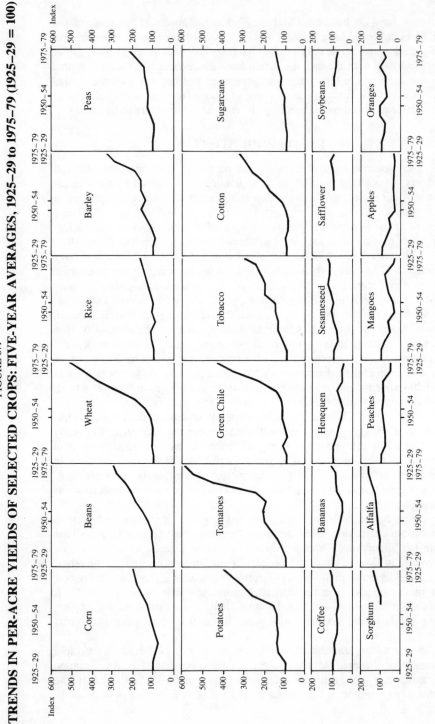

Source: Secretaría de Agricultura y Recursos Hidráulicos, Dirección General de Economía Agrícola, *Consumos Aparentes de Productos Agropecuarios, 1925–1976,* and subsequent annual reports of Dept. of Agriculture, Bureau of Agricultural Economics.

higher demand the new land may be less favorable than the old, so that yields decline. However, in Mexico this factor does not appear to have applied universally: potatoes, chiles, and tomatoes have all registered enormous area expansion and simultaneously substantial increases in yields.

Another noteworthy characteristic is that almost all the major improvements in crop yields occurred in the period after 1955. It would seem that the impact of technological progress did not begin to show up in better yields until the mid-fifties, some time after the major new irrigation works had become well-established. It also appears that during the seventies the rate of growth in yields of several crops has shown a tendency to slacken off, though it would be premature to conclude that this implies establishing a more modest trend line. As already observed, the technical efficiency in many irrigation districts is now high, and future gains can be made only more gradually; while on the poorer of the rain-fed lands the scope for improvement remains, as hitherto, severely limited.

There can be no doubt that over the past three decades the expansion of irrigation works has been an important factor in raising the productivity of crop land. Especially in certain crops the proportion grown under irrigation has increased markedly (for instance, wheat, barley, potatoes, and tobacco), and this change can be correlated with rapid yield increases in each case. However, also in the case of crops always grown under irrigation, such as cotton, yields have improved significantly, which is but a reminder that other factors have importance too: seed varieties, pest control, rotations, soil management, and so on.

It is also useful to consider the overall productivity of crop land in terms of the value of its output. Using annual Department of Agriculture data and three-year averages to reduce the weather effect, it appears that at constant prices the average per-acre value of output has slightly more than doubled between 1950 and 1980, with most of the increase occurring between 1950 and 1965. Over the whole period the cumulative rate of growth was 2.3 percent. Four states of the northwest achieved rates superior to three percent, as also did Guanajuato and Jalisco, while 16 states of the center and south recorded rates below two percent. Significantly, the fastest growth in productivity was registered in those states which most benefitted by irrigation, which incidentally permitted a certain shift to higher value crops in some districts (for example to tomatoes or to strawberries).

The performance over this period by the major groups of farmers was quite contrasted. (For this purpose we have to use census data since the annual figures do not give breakdowns by tenure, and so we cannot go beyond 1970.) As can be seen in Table 6.4, the larger farms increased their per-acre productivity by 147 percent during the 1950–70 period, the small ones by 73 percent and the ejidatarios and comuneros by 113 percent. Expressing the evolution in another way: the ejidatarios' productivity in 1950 averaged 77 percent of that of the larger private farms, but only 67 percent in 1970. These figures do not show contrasts as large as some authors have suggested, but they are subject to the reservation that much ejido land is being (illegally) cultivated by private farmers.

TABLE 6.4

VALUE OF CROP OUTPUT PER ARABLE ACRE

(pesos per acre at 1970 prices)

	1950	1970	Increase (1950 = 100)
Private Farms			
over 12.5 acres	238	589	247
under 12.5 acres	298	514	173
Ejidos	185	393	213

SOURCE: Statistical Office, decennial farm censuses.

NOTE: Communities in 1950 included with private farms; in 1970 with ejidos.

Physical productivity in farming is not solely a matter of crops and their yields; it also includes grasslands and livestock. Since these aspects have been analyzed in the preceding chapter, it will be sufficient here to recapitulate the main conclusions. As for pastures, two opposing trends were observed: first, an increase in productivity due to the expansion of the area of cultivated grasses, and secondly a general deterioration of the natural grasslands through overstocking and neglect. One can reasonably expect that in the future the positive balance between these two trends will be retained.

As for livestock, which is such an obscure topic in Mexico, explanations can be found in the previous chapter and in the Appendix as to how the volume of output and its growth have been calculated. Relating this output to the combined total of grass and crop land (since animals consume feed from both), and using 1970 prices, the value in real terms of livestock output per acre increased by 77 percent between 1950 and 1970, equivalent to a cumulative annual growth of 2.9 percent. Though complete data for the seventies are lacking, it seems that a similar rhythm of growth has been maintained. Between 1950 and 1970 the growth was composed almost equally of an increase in the numbers of livestock per acre (31.7 percent) and an increase in output per animal (34.6 percent).

There are wide differences in the productive quality of livestock between regions and between farms of different types of tenure. The evidence on this is unfortunately available only in the census which so grievously underestimates the total output, but probably this does not appreciably influence the relative performance of the different groups. Thus in Table 6.5, taken from the 1970 farm census, it can be seen that in the northwestern region the animal units were more than twice as productive as those in the two southern regions. We cannot be sure that these figures reflect the situation correctly. It might be that the regions in which the census was most defective in registering output of meat, milk, and eggs were those subsistence zones in which most of the production was for own use of farm families, and if this were so it would imply greater underestimation in the south than in the commercialized northwest.

Looking at the breakdown of animal productivity by farm tenure, we

TABLE 6.5

VALUE OF LIVESTOCK OUTPUT PER ANIMAL UNIT,
BY REGION AND TYPE OF TENURE: 1969

(pesos per animal unit)

Region	National Average	Ejidos & Communities	Private Farms	Private Farms' Performance (Ejidos & Communities = 100)
Northwest	844	354	1,081	305.4
North	624	325	792	243.7
Northeast	705	338	821	242.9
North-center	467	302	593	196.4
West-center	470	267	625	234.1
Center	730	218	1,192	546.8
Gulf-south	372	259	432	166.8
Peninsula	394	320	415	129.7
Mexico	558	284	721	253.9

SOURCE: Statistical Office, farm census of 1970.
NOTE: These output figures are not the author's estimates, but those recorded by the farm census. They include value of output and sales of live animals.

see that everywhere except in the two southern regions the animals were more than twice as productive on private farms as on those of ejidatarios and comuneros; in the northwest the recorded difference was more than threefold and in the center more than fivefold. Undoubtedly productivity differences between the two groups of farmers do exist, but it is not credible that they should be so large. Inasmuch as subsistence production is more widely spread among ejidatarios and comuneros, it is probable that under-reporting affected this group more severely than the other, but we have no basis for guessing by how much the ejidatarios' animal productivity should be raised. Of course, under-reporting also occurs on private farms, especially in respect to numbers of cattle, since ranchers are supposed to own not more than five hundred head. Regrettably the topic must be left in this inconclusive state.

By combining the output of crops with that of livestock products, we obtain figures of total agricultural output which, when related to the agricultural area, give some indications of productivity as a whole. However, in the case of Mexico certain conceptual difficulties are encountered. First, as noted earlier, the expansion of crop production has been much more rapid according to the annual figures of the Department of Agriculture than according to the censuses, and the divergence has been particularly marked since 1960. There are reasons for supposing that the true rate of growth lies somewhere between the two, but since at this writing we have to use for the seventies the crop figures of the department, we must constantly bear in mind the element of exaggeration which they seem to contain. Moreover, by using the department's figures we can employ three-year averages (thus reducing the factor of unrepresentative weather), which is impossible if census data are used. Secondly, an element of double counting enters in when we add crops and livestock products together, because we have not deducted the feed crops used by farm animals.

Thirdly, the distortion created by relating output to the entire agricultural area is much greater when crops are included along with livestock products in the output total, and for this reason productivity could also be calculated for crop land only. The former method of course reduces the apparent land productivity in regions where pasture areas predominate, whereas the latter method exaggerates the productivity in these same regions.

TABLE 6.6

COMBINED VALUE OF CROP AND LIVESTOCK PRODUCTION:
TOTAL AND PER ACRE OF CROP AND GRASSLAND, 1950 AND 1970

Region	Value of Production (in million pesos at 1970 prices)		Increase 1950 to 1970	Production per Acre of Crop and Grassland (pesos per acre at 1970 prices)		Increase 1950 to 1970
	1950	1970	(1950 = 100)	1950	1970	(1950 = 100)
Northwest	4,180	11,048	264	126.7	246.1	194
North	4,002	7,152	179	53.0	86.2	163
Northeast	2,182	4,289	197	166.7	232.7	140
North-center	1,830	3,671	201	87.4	172.0	197
West-center	5,055	10,591	210	165.9	330.6	199
Center	3,368	6,260	186	311.2	598.9	192
Gulf-south	4,466	9,820	220	203.6	365.4	179
Peninsula	561	857	153	124.2	196.7	158
Mexico	25,644	53,688	209	122.2	222.6	182

SOURCE: Production data from Table 1.5; area data from Table 3.1.

Using the first of the two methods, Table 6.6 shows that land productivity increased by some 82 percent between 1950 and 1970 or at an annual average rate of 3.1 percent. During the seventies the department's figures report a similar rate of growth, but the real figure is certainly below three percent. The regional growth rates were all remarkably close to one another, except the rather poorer performance of the north-central region and the peninsula. In absolute terms the per-acre output of the center region in 1970 was almost seven times that of the north, but this is where the distortion most shows itself. The center has relatively little pasture and much crop land whereas the northern states have vast areas of semidesert classified as pasture. Even the northwest region has much of this latter which pulls down its productivity classified in this way.

Measuring the other way, in terms of productivity of crop land only, the northwest because of its irrigation has the highest figure, more than 50 percent above the national average. The central and southern regions all come below the average. Since 1970 there are indications that productivity has grown slightly more rapidly in some of the central states than in those of the north and northwest, but the evidence is as yet inconclusive.

When it is recalled that between 1950 and 1970 total agricultural production rose at a rate of 3.8 percent per year in real terms, it is apparent that most of this was a productivity gain; the agricultural area increased only at 0.7 percent per year. This same phenomenon has almost certainly continued through the seventies, though data are not yet available.

What further gains in productivity is it legitimate and realistically probable for Mexico to expect to achieve during the 1980s? Inasmuch as the components of productivity are numerous and interrelated, it is desirable to consider the various factors separately and then together. For this purpose we may construct a simple model for 1980 to 1990 of the likely evolution of the productivity factors, as was done in Chapter Two for consumer demand.

Let us start with the area of farm land, as being the most basic component. In Chapter Three we surmised that the crop area could perhaps expand at about 0.4 percent per year while the grassland area might increase at 0.3 percent mainly through conversion of unproductive forest and scrub lands. Most of this increase is likely to occur in the southern regions which unfortunately are suited to only a limited range of crops.* This is a gloomy conclusion when it is recalled that population is expected to be rising at 2.7 percent and total food demand at something approaching five percent per year.

The prospects are a little brighter for increase in the *harvested* area, for reasons mentioned earlier. The amount of fallow land will continue to be reduced, crop losses through pests and diseases should diminish (though not much for those caused by floods and droughts), and finally the area under double cropping has potential for being expanded. As a result it should be feasible to increase the harvested area at a rate approaching one percent per year during this decade.

The next factor is crop yield per harvested acre. Reasons have been given for believing that the Department of Agriculture's published data for the 1950–70 period exaggerated the rate of growth of yields at 3.1 percent per year. But its figures for the seventies show a much lower figure of 1.5 percent. Possibly the performance in the eighties will be no better, yet the government has become more conscious of the shortfall in food production and may use some of the additional revenue generated by petroleum to accelerate technological progress in the farm sector. We therefore put in the model a rather optimistic yield growth rate of two percent. The combination of

---

*No allowance has been made for land that will be taken from agriculture for urbanization, since this will be quantitatively unimportant. Neither has allowance been made for marginal arable and pasture land which may be abandoned or converted to forest. To this extent the projection errs on the side of optimism.

harvested area and yield gives a three percent annual increase as our projection for crop production.

The increase in numbers of livestock during the fifties and sixties averaged 2.2 percent annually (calculated in animal units and using census data which appeared more reliable). The year to year figures from the Department of Agriculture during the seventies give a growth rate slightly below two percent, which seem most improbable in face of the strong consumer demand for livestock products. And since incomes can be expected to grow more rapidly in the eighties, an incentive should exist to expand this sector while increasing its dependence on imported feeds. Numbers could be expected to grow at something like three percent, unless some future administration fixes maximum prices for meat, milk, and eggs at unprofitable levels.

The product output per animal unit had been increasing at about 1.3 percent per year, but there are reasons for supposing that this rate is declining. A major factor in the improvement was the poultry industry which became fully commercial during the past thirty years and is approaching its productivity ceilings. Progress in other classes of livestock is inevitably slower, so the projection sets a figure of only one percent per annum increase in output per animal unit. However, this means that altogether animal production may be increasing at four percent per year.

The combined output of crop and stock products may thus be expected to expand at about 3.5 percent during the eighties provided all goes well. This would be a fractionally better showing than in the fifties and sixties, and a considerable improvement on the poor performance of the seventies. If it errs, it is on the side of optimism. Yet our projections for demand have indicated for the same period of the eighties an annual increase between 4.5 and five percent. That is why it seems inevitable that Mexico's food imports will be rising, though simultaneously she will continue exporting certain tropical commodities.

With these tentative projections for production prospects in the eighties the analysis of what may be termed the physical side of Mexico's agricultural dilemma is now completed. We have seen how during recent decades a very considerable mobilization of physical resources has been accomplished: of land, of water, of animals, and of technology. For a time the rate of growth of farm production was one of the fastest in the third world. From 1965 onward, the pace of expansion slackened while the demand for farm products continued to accelerate, and this discrepancy has continued throughout the seventies.

The physical reasons for this are quite easily identified. It has become apparent that henceforth the bringing of new land into cultivation will raise formidable agronomic and management problems, further extensions to the irrigated area will be costly for the most part, because the remaining possible sites are physically awkward to develop. Livestock expansion faces the obstacle of rundown pastures that will not sustain herd increases and high costs of domestically produced feeds. Research scored a brilliant breakthrough in wheat breeding but faces more intractable problems in other crops and has neglected livestock. The farmers most receptive to technology have already modernized their practices, and those that remain to be approached are deeply traditionalist.

Such findings are discouraging, but they do not add up to a situation of hopelessness. We have not yet considered the human factor, and throughout history it has been human effort which has transformed unpromising resources into wealth. In Mexico, as elsewhere, much depends upon the motivation of the farm people, and this in turn depends on the institutions within which they operate, on the economic incentives offered by the market, or by government policies toward farming, as well as on the prevailing social and political environment. In the following chapters we turn to these human, economic, and social matters.

7. Farm People

In order to understand the why and wherefore of the present stratification of Mexico's farm population, we have to be acquainted with the salient facts of the Agrarian Reform. At the time of the Mexican Revolution (1910) the farm land was almost entirely in the hands of a very few thousand land-owners, some of whom owned several million acres each. They were either the descendants of Spanish gentry from colonial times or they (including a number of foreigners) had been granted concessions during the regime of Porfirio Díaz.

A limited number of them engaged in ranching while others devoted a portion of their estates to the cultivation of crops. In a few districts of the north and northwest, American enterprises had obtained concessions for constructing irrigation works and were growing cotton and sugarcane. To a large extent, however, the lands of the *hacendados* were left idle. Down below were the landless laborers who sought work on the estates; but the volume of employment was insufficient and the pay was atrocious, so that rural poverty was grim and widespread.

The grotesque maldistribution of land ownership, coupled with the arrogant behavior of many hacendados enabled Zapata to arouse a tremendous response to his cry: "The land belongs to those who work it," while the intellectuals who were formulating the outlines of an agrarian reform, rightly identified landlessness with insecurity and poverty. The aim of these latter was to inaugurate a process of redistribution which would make land available to all who wanted to cultivate it, and in this way lay a firm foundation for security and prosperity in the countryside.

This seemed an entirely feasible objective. The number of families at that time dependent on agriculture was estimated to be something less than two million, and although the amount of arable land (or more precisely potentially arable land, because much good land was left uncultivated by its owners) was unknown because no agricultural census had ever been taken, it seemed that out of the 487 million acres which comprised the area of the Mexican Republic, there must surely be enough and more than enough to satisfy the wishes of farm people.

The first Agrarian Reform Act was promulgated in 1915, and its main features were subsequently incorporated in Article 27 of the Mexican Constitution of 1917. Since that time there have been modifications and additions, the latest and currently operative text being that contained in the Act of 1971. The original Act of 1915 gave the government authority to expropriate parts of the large estates, subject to compensation, and to distribute this land to applicants who satisfied certain conditions. It was thus concerned principally with the process of physical redistribution. In marked contrast to agrarian reforms which have been effected in developing countries in more recent times, the Act prescribed no limits as to the amount of land to be distributed nor as to the time within which the process should be completed.

This latter was at once a strength and a weakness. It was an advantage in that it allowed land redistribution to proceed slowly and so avoid dislocating the nation's food production. In fact it was not until the thirties, after the Act had been law for twenty years, that redistribution on a substantial scale was undertaken. Then the activity died down during the forties and fifties, to become vigorous again in the sixties and seventies.

The extreme gradualness of the execution of the reform program had, however, its negative aspect because, as a consequence, the remaining large landowners never knew when their turn would come. Most of them had no titles and no security of tenure. Facing uncertainty, they were loath to invest in land improvement operations, especially since the Act's compensation provisions in the event of expropriation had never been implemented.

The redistribution program granted land to two distinct categories of persons. The first of these were the agricultural communities ("comunidades"), village-based organizations which had held their land in common in ancient times but had lost out in the mid-nineteenth century when liberalistic governments granted individual land ownership to each community member who thereupon sold it (maybe under pressure) to a neighboring landowner. It was these communities whose interests Zapata championed, and the first task of the Agrarian Reform was to restore to them their lands. This was not fully possible: written records had been lost or had never existed. Some communities were never reestablished or reemerged as ejidos. Others were newly invented, while not a few ended up with more land than they had held in the old days. This was a minor aspect of the Reform. In the country as a whole, there have never been as many as 2,000 communities, with some 200,000 members and occupying some 22 million acres, which is less than seven percent of the total agricultural area.

The second task of the Agrarian Reform, and in the long run far the more important one, was the creation of ejidos, which in the course of the succeeding sixty years came into possession of over 40 percent of the agricultural land of the country. To establish an ejido, a group of twenty or more persons living in a village (later the condition of geographical propinquity was waived) made application for a land grant. After checking the credentials of the group and its members, a piece of land was assigned to the "ejido" as a legal entity and as legal owner. The ejido assigned a plot, usually about 25 acres, to each member who enjoyed rights of occupancy for himself and his heirs.

This system of land tenure was supposed to recreate in part the pre-Columbian tenure arrangements of the Aztecs and in part the system which had prevailed in medieval Spain—hence the name "ejido." Its mode of operation will be examined in detail presently. Suffice to say here that the ejido system has become the principal, and most hotly discussed, feature of Mexico's agricultural institutions; nothing quite like it exists in any other country.

Meanwhile, a third class of farm operators was coming into existence. It did not take the hacendados long to realize that, faced with the prospect of expropriation without compensation, the wise thing to do was to sell portions of their land to would-be farmers whenever opportunity offered. This they did, and thus was created a new group called the "small owners" (*pequeños propietarios*). The group was composed partly of the purchasers just referred to, partly of descendants of the hacendados still operating the vestigial remains of their estates, and partly of "colonists," a special legal status devised for persons to whom individually the government assigned lands in new areas opened up for settlement.

The members of this group now occupy nearly half the agricultural land of the country. Many of them are indeed small-scale farmers, but others are small only in name and may own (legally) up to 750 acres of plantation crops or alternatively several thousand acres of grassland, which is why it seems better to call them "private farmers," as distinct from the communal and ejidal farmers neither of whom own land individually.

In the early years of the reform it was imagined that there would be enough land for all applicants, and that it would be sufficient to expropriate only the unused parts of the large estates, leaving their owners' genuine agricultural activities virtually untouched. But the subsequent demographic explosion created a land shortage; the estates had to be dismembered more and more. In 1946 the then government wrote into the Reform Act and into the Constitution an amendment prescribing upper limits to the amounts of land and cattle that any individual might own—an amendment which has created innumerable problems which will be discussed later. Even so, there are still vast numbers of applicants, probably as many as there were sixty years ago, but there is no land left to distribute except by (a) chasing the farmers who evade the law and illegally operate large units, (b) dispossessing ejidatarios who fail to cultivate their lands and (c) opening new lands, e.g., in the southeast, which are automatically assigned to ejidos.

Those people who view the problem from the angle of land availability insist that the Agrarian Reform as a redistribution operation has been completed, and urge the government to declare the fact publicly. Those who take the side of the people who would still like to obtain farms argue that the Agrarian Reform can never be terminated until the last claimant has been satisfied. These irreconcilable positions underlie much of the present rural discontent and uncertainty.

QUANTIFYING THE FARM POPULATION

How many people are occupied in agriculture in Mexico? This seems a silly question, because it is too easy. For the United States such information is readily available in the censuses, and any discrepancies between one census and another can be explained from identifiable differences in concepts and definitions. Not so in Mexico. There are two principal sources of information, namely the population census and the agricultural census. The latest published figures refer to 1970, when each had its last count, and while the population census reported 5.1 million persons engaged in agriculture, the agricultural one reported 7.8 million. Between 1940 and 1970 the former recorded an increase of 33 percent and the latter of 66 percent. These are such wide differences that we are obliged to devote some space to trying to explain them, even at the risk of being a little tedious; otherwise we would be precluded from elaborating analyses and drawing conclusions.

The conflicts in the evidence derive only to a minor degree from deficiencies in the collection of the basic data. Most of them arise from ambiguities and confusions in the definitions and concepts. For instance, what do we mean by "a person engaged in agriculture"? We may assume that the definition would embrace the cultivators of corn and beans, even though they actually work only at sowing and at harvest time, remaining idle or having other occupations during nine months of the year. But what about the farmer's wife who may spend a few minutes a day feeding the chickens and rabbits but devotes the rest of her time to housekeeping? What about his children who may lend a hand at harvest?

Then there are the seasonal workers, those who go cotton picking, cane cutting, coffee harvesting, etc. Some are casual laborers who move from district to district in search of work and may even be found working on building sites when there is no farm work. Some are farmers or the sons of farmers who are able to undertake additional work either because it is offered at a time of year when there is little activity on their own land, or because their farms are too small to provide full-time employment for all the family.

In countries in which subsistence farming has virtually disappeared and where agriculture has become a commercial operation, these phenomena of semi-employment and dual employment no longer exist. But in a country like Mexico they are more frequent than straightforward full-time agricultural work; and so definitions have to be elaborated. For example, if a man has two occupations, it is the one to which he devotes most time that is recorded in the statistics. However, if a man has only one occupation, it is so registered no matter how little time he dedicates to it. In the cases of women and children it is usual to require an input of a certain number of hours per week or per year to qualify them as engaged in an occupation.

Such definitions have not yet been systematically formulated in Mexico, and the few that have may not be adhered to in practice. Thus, if an ejidatario spent most of his time in bricklaying, he would be unlikely to admit it, for fear of losing his rights in his ejido. And in regard to ejidatarios there is the further problem of legal status. In 1970 the agricultural census, using a relatively loose definition, recorded 2,012,856 ejidatarios, while the population census using a tighter definition enumerated only 814,006. Quite a number of persons are members of ejidos but have no plot of land (or "parcela," as it is called), because there is none yet available or for other reasons; and these have no agricultural activity. On the other hand, a considerable number of ejidatarios cultivate their parcelas and in addition have bought and cultivate other pieces of land, being thus at one and the same time ejidatarios and private farmers. One has to be careful not to count them twice. The same caution applies to the numerous ejidatarios who, though registered in this category, spend most of their time as hired farm workers.

It is hardly surprising that the Mexican census takers have not yet succeeded in overcoming these problems. For instance, between 1960 and 1970 the number of family workers recorded by the population census rose from 159,000 to 528,000, while during the same period according to the agricultural census it fell from 3.3 to 2.1 million. Likewise, according to the former census, the number of hired workers fell from 3.2 to 2.5 million, while according to the other source it rose from 1.6 to 2.5 million, a fortuitous agreement this last because in individual states the two sources differed greatly.

We now come to a feature which gives us an important clue to unravelling the statistical confusions, namely the seasonality of farm work. In all parts of the world the cycle of crop growth imposes a seasonal rhythm on farmers' activities, a rhythm which is most pronounced in regions where monoculture prevails. To some extent this seasonality can be reduced through crop diversification, and especially through double cropping, where irrigation makes this practice possible, but the chief factor for levelling the work load is the presence of livestock.

In countries like Mexico the seasonal distribution of farm work is extremely pronounced, partly because animal husbandry remains unimportant, partly because climate imposes limitations on crop diversification, and partly because relatively few farmers have reached the degree of mechanization which greatly reduces labor requirements at peak periods. Mexico has two main harvest periods, one for winter crops and the other for summer ones. According to the 1970 census, just over six million persons were registered at the peak moment for harvesting winter crops, and 9.7 million at the summer peak. No count was taken of numbers in the least active season of the year; it was certainly much less than six million.

It is this phenomenon which enables different authors to put forward such different assertions as to the size of the farm population. Depending on what political axe they have to grind, some will say "nearly 10 million," others will quote six and yet others perhaps three million. Each of them will be right in respect to a particular moment in time. This problem of seasonality, together with those of double counting and of dual occupations already mentioned, makes it virtually impossible to arrive at a reliable figure if the approach is by individuals.

In Mexico a more realistic approach is to try to estimate the number of farm families, defining these as families which obtain more than half their income from farming activities, irrespective of whether those activities occurred on their own farms or on the farms of others, or a combination of the two.

Let us begin with the ejidatarios, of whom the 1970 census recorded two million, in round numbers. Evidence is available to suggest that some 200,000 of these were engaged primarily in nonagricultural activities. A further 200,000 were also owners of private farms; it does not matter whether we list them as ejidatarios or as private farmers, but we must not count them twice. We are left, therefore, with 1.6 million full-time ejidatarios.

The members of communities numbered 200,000, and, since we have no information about their nonfarm activities, we will list all of them as farm families.

Turning now to the private farmers, the 1970 census recorded just over 900,000 private units having farm land, and we have to estimate how many of these were operated by persons having their principal occupation outside agriculture. We saw that among ejidatarios the proportion was about 10 percent, and we know that this practice of combining jobs is much more frequent among persons whose farm land is privately owned; so let us guess the figure at one-third of the total, or 300,000. Support for this view comes from the fact that of the private farms some 500,000 were under 12.5 acres in size, and the majority of these would be part-time units. This leaves 600,000 farm families full-time in the private sector.

Among the hired laborers, the 1970 census distinguished 434,000 as working full-time on farms. A few of these may also have land of their own and have been already counted, so we may round the figure down to 400,000. Finally, there are the seasonal workers who, although not engaged full-time, have no other occupation and therefore "depend on agriculture." Evidence from the Bank of Mexico's incomes study of 1968 suggests that these may number about 500,000. (To some this may appear a low figure, but the great majority of seasonal workers are members of the families of farm operators already listed above.) The total picture appears in Table 7.1.

Altogether we estimate that in 1970 there were 3.3 million families dependent on farming, and of these some 2.4 million were farm operators. These, as the reader will appreciate, are approximate calculations, but probably the best that can be made after sorting the conflicting evidence. The overall situation was probably little different in 1980, because a probable increase in the number of ejidatarios will have been offset by a decline in the number of private farmers.

It remains to be asked how this figure of 3.3 million families can be reconciled with the much higher figures quoted earlier for the peak harvest periods. For this exercise, what we need to do is to visualize those 3.3 million families as one gigantic sea anemone which contracts and expands, not in this case according to the temperature of the water, but in response to the manpower requirements of agriculture at the different seasons of the year. In the

TABLE 7.1
ESTIMATE OF NUMBER OF FAMILIES
MAINLY DEPENDENT ON FARMING: 1960 AND 1970
(in thousand families)

	1960	1970
Ejidatarios, total	1,593	2,012
Ejidatarios with mainly		
nonfarming activities	151	200
Full-time ejidatarios	1,447	1,800
Ejidatarios also having		
private farms	288	200
Full-time ejidatarios		
without private farms	1,159	1,600
Comuneros	200	200
Private farmers, total	1,146	900
Private farmers with mainly		
nonfarming activities	382	300
Full-time private farmers	764	600
Full-time hired workers	. . .	400
Seasonal workers dependent		
mainly on farm wages	. . .	500
Hired workers, total	900	900
Total	3,023	3,300

SOURCES: Farm censuses and author's estimates (see text).

dormant season, when very little labor is needed, there may be less than one job per family, reflected in the fact that even the head of the family sits around doing nothing and waiting for his corn to grow. If this slack period occurred in the same month all over the country, which it does not, and if we took a census at that moment, we would probably find much less than three million people at work.

Then the field activities increase, the farmer works full-time, his son helps him, some of the landless families obtain temporary employment, so that the total number at work rises to five million, to six million, and so on. Finally at the peak moment of labor requirements, all the members of the family, who are not too old or too young, are out in the fields, all the landless laborers find employment though not for very long, and at this moment the economically active agricultural population may easily rise to 10 million. After the summer harvest the manpower requirement declines suddenly to a low figure and the cycle recommences. We now recognize the pitfalls that occur in trying to calculate the farm population in terms of numbers of individuals, and why it is more realistic to enumerate families whose effective contribution in manhours varies enormously from one season to another.

Before leaving the subject of the farm population a few comments are required on its geographical distribution. The reader should visualize the northern half of Mexico as essentially cattle country—wide open spaces characterized by low rainfall and poor quality grazing. Here and there one finds irrigation and intensive farming, but this occupies but a small fraction of the total area. This northern half accommodates only one quarter of the farm population. The remainder lives in the southern half, not because of its superior agricultural productivity, since except in the coastal belts most of the soils are eroded and of low fertility, but because these areas have for centuries been the home of the indigenous inhabitants. So long as the population was small, the man/land ratio in these areas continued to be tolerable; but with the fall in mortality and the recent rapid increase in numbers, the southern half has become seriously overpopulated.

One sign of the overwhelmingly agricultural character of the center and south can be seen in the ratio of farm population to total population, a ratio which exceeds 70 percent in almost all the states of these regions, and even exceeds 80 percent in Hidalgo, Tlaxcala, and Yucatán. By contrast, in Baja California Norte the ratio is under 10 percent and Nuevo León 17 percent.

To some extent the population pressure has been and is being relieved by internal and external migration. During the forties and fifties substantial numbers migrated from the central states to Sonora and Sinaloa to take advantage of the newly developed irrigation. Since then the main employment opportunities have been urban ones, first and foremost in Mexico City and latterly in Monterrey and Guadalajara as well. Thus, between 1960 and 1970 the farm population declined in absolute terms in the Federal District and in the states of Puebla and México, both of them adjacent to the capital; it also declined in Nuevo Leon whose capital is Monterrey, and increased only three percent in Jalisco (capital: Guadalajara). No doubt if a larger number of provincial cities could acquire the dynamism of those just mentioned—the policy of President López Portillo sought to induce such growth—the underemployment in the agricultural sector could be relieved in other regions too.

There are a number of people in Mexico, as indeed in other countries, who while welcoming an expansion in urban employment would deprecate a decline in the farm population. For them the occupation of farming offers a way of life superior to that of urban-based occupations, and they would regret seeing the farm sector lose its relative importance. This represents of course a metaphysical attitude to the structure of human society. Nonetheless, desirable as such an Arcadia might be in theory, historical experience has shown that a progressive division of labor offers the only road to achieving acceptable standards of health, education, and culture. Moreover, a farm family cannot enjoy these benefits if it is producing food for only itself and two or three other families; it must produce enough for thirty or forty families. Mexico is moving in this direction: the farm population is declining in percentage terms though not yet in absolute numbers. The future prospects in this respect will be reviewed in Chapter 10.

THE EJIDATARIOS

From a review of the farm population as a whole, we now turn to consider the principal groups of producers, beginning with the ejidatarios who, at over two million, are numerically by far the most important.

At the last count in 1970 there were over 21,000 ejidos, and since then the number has slightly increased. This implies that the average ejido has somewhat less than one hundred members. The average size would be around 7,500 acres, of which 1,250 would be cropland, 3,000 pasture and 3,250 forest and waste areas. Nevertheless there are some extremely large ejidos of 750,000 acres or more, and with more than 1,000 members. At the other extreme are some mini-ejidos, of less than 250 acres, with ten or a dozen members.

When a grant of land is made to a group of people who have organized themselves into an ejido, the Assembly decides what is the area of actual or potential cropland, and this area is divided into plots or "parcelas," one per member. The remaining lands, that is the pasture, woodland, and other areas remain in the communal ownership of the ejido, though members have the right to graze their animals on the pasture and to cut firewood in the forest. Some ejidos when they were founded obtained relatively generous land grants and were able to allot as much as 50 acres of cropland to each member; the least fortunate may have allotted only 10 or 12 acres; the norm was supposed to be 25 acres. Inasmuch as most ejidatarios have several sons, the parcelas have tended to be subdivided (though this is illegal), and now one third of the 1.8 million parcelas are smaller than five acres while nearly another third are between five and 12.5 acres. Less than four percent are larger than 50 acres.

Once an ejido has been established, the pattern of land occupancy becomes frozen. A member may permit each of his sons to operate a part of his parcela, while remaining himself nominally the occupant, but he cannot enlarge his farm by taking over someone else's parcela. His only recourse is to save enough money to rent or purchase a piece of private sector land; and usually there is none available within a convenient distance. This makes it almost impossible for the efficient ejidatario—and there are efficient ones— to enlarge his scale of operations.

On the other hand, an ejidatario who has no taste for farming or whose sons have moved into other occupations can neither rent his parcela nor sell it. If he wishes to leave, all he can do is to surrender it to the ejido which will assign it to someone on the waiting list but will give him no compensation for any capital improvements he may have made. Hence there exists an unknown but certainly large number of ejidatarios who are bad and unwilling farmers, but have no way of extricating themselves from their situation. In the more fertile regions, it is true (and especially in the irrigation districts), many ejidatarios, in some districts more than half of them, rent their parcelas illegally to private farmers and live off their rents; which is one way for private farmers to evade the law regarding maximum size of farm.

These restrictions on the ejidatarios have their justification in the context of their historical origin. When the designers of the Agrarian Reform were engaged in their task, the triumph of the Revolution over the hacendados was by no means assured. They resisted every step in the Reform, and they employed their White Guards to resist with violence. There was real danger that they would exploit the ignorance and weak bargaining position of the new ejidatarios to buy back much of the land that was being distributed. It was therefore essential to give legal protection to the ejidatarios by means of these restrictions on renting and selling. Without this, the democratic pattern of land holding could not have been achieved or maintained.

But while this and other provisions in the early agrarian legislation adequately met the needs of the time, it is less certain that they are well suited to the needs of the 1980s. The agrarian reformers, just like reformers in so many other moments of human history, established a sytem as rigid as the one they overthrew; they appear to have underestimated the strength of the winds of change which were already stirring in Mexico: demographic change, technological change, social change. For instance, it was probably insufficiently realized by persons accustomed to the static agriculture of past centuries, how much a farmer can improve the productivity of his land by investment and hard work, because at that time the power of the application of technology and the role of human ingenuity had not become apparent. So no provision was made in the legislation either to encourage ejidatarios to improve their land or to compensate them for improvements if they wished to renounce their parcelas.

A farmer will improve his land if he has security of tenure or if he has legal entitlement to compensation when his occupancy is terminated. The ejidatario has neither. Theoretically after he enters into his parcela and receives a *certificado de derecho agrario* (certificate of agrarian rights), the ejido should proceed to survey the land and establish boundaries to the parcelas, after which each ejidatario should receive a *titulo de propiedad parcelaria* (parcela title). But partly because the ejidatario would be required to pay the land survey costs, and more importantly because the legislation prohibits the parcellation of ejidos if the resulting units would be below the legal minimum size (and the great majority would be below!), the result is that only about seven percent of the ejidatarios have acquired titles. The remainder cultivate parcelas whose boundaries are undetermined except by oral tradition, and they remain at the mercy of the ejido president or other local boss who may requisition their land for his friends, assigning them in exchange parcelas of much inferior quality.

Mention was made of the land held communally by the ejido. As to the forest, very few ejidos exploit it themselves, and much of what they own is of little commercial value. The better forest lands they rent out to private enterprise concessionaires. But the ejidos own one third of the nation's grasslands. Unfortunately, the reformers did not envisage that the progress of technology would require investment in pasture improvement—reseeding, fertilizing, fencing, elimination of bushes and noxious plants, provision of watering facilities and so on. All this theoretically is the responsibility of the ejido, but very little such investment is ever undertaken, because the ejido is not organized for such activities and because the members, not seeing any personal advantage resulting to them from such expenditure, would be unwilling to make financial contributions. Consequently, since the system is one which fails to stimulate the application of technology, almost all ejido pastures are badly deteriorated.

THE COMUNEROS

The information on the agricultural communities is conflicting, largely because many census enumerators are not clear about the difference between a community and an ejido. Thus, the recorded number of communities varies greatly from census to census, although we know that they are in reality extremely stable institutions whose numbers do not change. The 1970 census recorded 1,231 communities having 205,616 members and owning 22.7 million acres. The average community therefore had 167 members and 18,440 acres, somewhat larger than the average ejido on both counts.

Although a few communities can be encountered in the developed districts, the great majority are located in the more mountainous and remote areas where they have little contact with the market economy, and in many instances their members speak indigenous languages rather than Spanish. As much as half their area is forest and wasteland, 40 percent is pasture and only 10 percent under crops. The largest concentrations of communities are found in Oaxaca and Guerrero.

Only two-thirds of the comuneros possess individual parcelas, the remainder being in collectives or some mixed regime. Since 1934 the legislation has recognized the right of individual comuneros to have parcelas up to 125 acres in size, provided the occupiers have been in possession for at least ten years. Otherwise the law treats the parcelas of comuneros in the same manner as those of ejidatarios with prohibition of renting and selling. In practice the majority of the communities permit their members to buy and sell land both within and outside the community's limits.

Because so much community territory is forest land, communally owned, naturally many conflicts have arisen concerning its exploitation. Strong arm tactics may be used by local bosses to cut more than their entitlements, or quarrels may develop between communities because the boundaries of their forests are ill-defined. Concessions have been granted to resin-tappers, in some cases on a considerable scale, with the result that after a few years the conceded forests have become virtually private property. Resin-tapping methods are generally primitive, and, in the absence of effective controls, the tappers are prone to exhaust rather than conserve this valuable resource.

Unlike the ejido which is a modern and artificial creation, the communities are based on long-standing local traditions. However, the efforts now being made by the authorities to improve the well-being of the indigenous groups, teaching them new crafts, diversifying their job opportunities and gradually incorporating them into Mexican society, may be expected slowly to weaken this form of land tenure until it is replaced either by collective organization of farming or by individual family ownership.

THE PRIVATE FARMERS

The private farmers are a much less homogeneous group of farm operators. Already the distribution by size of farm provides an indication. Thus of the 900,000 units recorded as belonging to this sector in 1970, well over 500,000 were below 12.5 acres in size, 200,000 had from 12.5 to 62.5 acres and less than 200,000 were over 62.5 acres. The majority of the mini-farms (those under 12.5 acres) are almost certainly part-time enterprises. Some 70 percent of these very small farms are located in the states of Hidalgo, México, Tlaxcala, Puebla, and Oaxaca. In recent times there has occurred a sharp reduction in the number of these units as their owners have sold out and migrated to the cities, (whereas the ejidatarios on equally small units, being prohibited from selling, are much less mobile).

A very distinct group is that of the private farmers in the irrigation districts. Although they comprise hardly more than 100,000 operators, they farm nearly 3.7 million acres of irrigated land, mostly in units of over 50 acres. (Their average unit size is almost 37 acres compared with 12 acres for ejidatarios in the same irrigation districts.) They are located principally in Sonora and Sinaloa and, to a lesser extent, the other northern states. They are a business-oriented group, employing the most modern techniques and obtaining high returns on their capital investment.

A third group, less organized and less powerful but nonetheless important, consists of the plantation owners of the coastal areas and the south. Their chief crops are coffee, cacao, bananas, coconut, and oil palms; some also grow sugarcane. Their operational practices tend to be more traditional; with some exceptions they have devoted less attention to organization and marketing.

Of extreme importance, the fourth group is that of the cattlemen. Some are to be found in the relatively new cattle raising areas of the south, but traditionally they ranch the open spaces of the north, many of them owning thousands of acres of grazing, albeit of poor quality. The majority specialize in raising steers for export to the U.S.A. They are highly organized and politically influential.

Among all these groups of private farmers, owner-occupancy predominates by more than 90 percent. In the whole Republic there are only some 25,000 share croppers and about 27,000 farmers renting the land they farm. However, a considerable number of those who own their homesteads also rent additional pieces of land. This arrangement is particularly common in the irrigation districts where, as already mentioned, private farmers illegally rent irrigated parcels from ejidatarios; for obvious reasons, the extent of the practice cannot be quantified.

The private farmers face a number of special problems. For instance, the mini-farmers in this sector lack the benefit of any advice from extensionists, because the official extension service is oriented exclusively to the ejidatarios and their needs. Likewise they lack farm credit (except of the extortionist kind obtainable through village moneylenders) because their enterprises are too small to interest the private banks while the official Rural

Credit Bank deals mainly with ejidatarios and otherwise with the larger private farmers. In the matter of marketing there are few cooperative organizations to help them dispose of their produce, though this disability they also share with the ejidatarios.

The larger-scale private farmers, on the other hand, can easily obtain credit from a variety of banks, provided the farmers have documented titles to their land (which many do not). They also obtain extension services through their farm organizations. Nonetheless, many are operating less efficiently than they could be, on account of the legal prohibition of mixed farming, of running a livestock operation alongside crop cultivation, or vice versa.

Worst of all, most of these farmers live in a continual state of uncertainty and insecurity. Because the articles of the Land Reform Act are so ambiguously drafted they do not know whether or not they are operating in contravention of the law. At any time they may be confronted with an organized invasion of several dozen or several hundred squatters, often well-armed. In such incidents neither the police nor the army will intervene. Paradoxically, those who practice the most flagrant evasion of the regulations regarding size of farm are the ones who have least to fear; they are the politicians or union leaders, local and national, whose activities cannot be questioned.

Nor is it solely the large-scale farmers and ranchers who are vulnerable. In recent years, many instances have occurred of invasions of small farms, say between 25 and 50 acres, often as a result of feuds between local families or political bosses. The invaders will camp on the properties, perhaps for weeks, until their claims to the land (all or a portion of it) are satisfied. Meanwhile, they may deliberately destroy crops, fencing, and livestock. Should the owner offer any resistance, he commits a breach of the peace, and for such an offense his certificate of inalienability may be cancelled.

Some invasions are believed orchestrated by officials of the Ministry of Agrarian Reform, and it is these persons, not the courts of law, who decide whether a property should be expropriated. In many instances, expropriation occurs neither because the farm owner is in breach of the law nor because some local landless laborers are asking for plots. However, in some cases of genuine local need, the Agrarian Reform Department may bring pressure to bear on some local landowners to sell portions of their properties to the government. For them such a solution is obviously preferable to seizure by force.

This atmosphere of insecurity which prevails in the Mexican farm community, not only in the northern states but also in such states as Colima, Chiapas and Veracruz, must be reckoned one of the principal causes of the stagnation of agricultural production since the sixties. It is true that those farmers who have already modernized their enterprises continue to maintain a high level of efficiency, maintaining a low political profile in order to keep out of trouble. But there are many, especially cattlemen, who could by long-term investment greatly improve the productivity of their properties. Insecurity of tenure deters them from doing so. It is impossible to calculate how much food production is being lost through this state of affairs, but it must be considerable because this group, cattlemen and crop farmers, accounts for the greater part of the nation's supplies.

REDISTRIBUTION AND FRAGMENTATION

Before examining the production performance of the various sections of the farm population, it is necessary to consider briefly what have been the main trends in land distribution and fragmentation. Such an analysis will assist in evaluating what are likely to be the future developments in regard to farm sizes and land tenure.

The following selected figures indicate in summary form the extent of land redistribution between the first and the fifth agricultural censuses:

	1930	*1970*
	(in million acres)	
Ejidos	20	150
Communities	15	23
Private farmers	290	173

From these figures can be seen the magnitude of the transfer of private farm land to ejidos. (The overall totals differ because a rather larger area was covered by the census at the latter date.)

There were two periods of intense agrarian reform activity, one in the thirties and another in the sixties. During the thirties there existed two sources of land for allocation to ejidos: one was the expropriation of parts of the many big estates which still persisted, and the other was the government programs of land clearance and irrigation works which brought several millions of additional acres into cultivation. By the sixties, however, this second source had almost dried up and almost all the transfers had to be at the expense of private landowners.

Also another remarkable change occurred. During the thirties, of the total land donated to ejidos some 23 percent was arable, somewhat higher than the national average percentage; but in the sixties only 11 percent was arable while half of the area transferred was forest and wasteland (Table 7.2).

TABLE 7.2

DECENNIAL INCREMENTS TO EJIDAL LANDS

	1930 to 1940	1940 to 1950	1950 to 1960	1960 to 1970	1930 to 1970
	Thousand Acres				
Arable	11,866	3,101	2,382	4,571	21,920
Pasture	17,549	14,507	7,621	15,471	55,148
Other	21,433	7,030	3,842	19,583	51,888
Total	50,848	24,638	13,845	39,625	128,956
	Percentage Distribution				
Arable	23.3	12.6	17.2	11.5	17.0
Pasture	34.5	58.9	55.0	39.0	42.8
Other	42.2	28.5	27.8	49.4	40.2
Total	100.0	100.0	100.0	100.0	100.0

SOURCE: Statistical Office, decennial farm censuses.

TABLE 7.3
LAND IN FARMS BY TYPE OF TENURE: 1930 TO 1970

Year	Thousand Acres				Percentage Distribution			
	Ejidos	Commu- nities	Private farms	Total	Ejidos	Commu- nities	Private farms	Total
				total area				
1930	20,620	14,826	289,724	325,170	6.3	4.6	89.1	100.0
1940	71,469	14,996	231,674	318,139	22.5	4.7	72.8	100.0
1950	96,107	18,666	244,799	359,572	26.7	5.2	68.1	100.0
1960	109,952	21,584	286,270	417,806	26.3	5.2	68.5	100.0
1970	149,577	22,711	173,326	345,614	43.3	6.6	50.1	100.0
				total arable				
1930	4,337	28,078		32,415	13.4	86.6		100.0
1940	16,202	19,061		35,263	45.9	54.1		100.0
1950	19,303	24,132		43,435	44.4	55.6		100.0
1960	21,685	28,024		49,709	43.6	56.4		100.0
1970	26,257	21,033		47,290	55.5	44.5		100.0
				irrigated arable				
1930	541	3,603		4,144	13.0	87.0		100.0
1940	2,457	2,059		4,516	54.4	45.6		100.0
1950	3,014	3,172		6,186	48.7	51.3		100.0
1960	3,529	5,158		8,686	40.6	59.4		100.0
1970	4,349	4,504		8,853	49.1	50.9		100.0
				grassland				
1930	8,789	155,515		164,304	5.3	94.7		100.0
1940	26,338	112,463		138,801	19.0	81.0		100.0
1950	40,846	125,648		166,494	24.5	75.5		100.0
1960	48,466	146,970		195,436	24.2	75.8		100.0
1970	63,937	130,034		193,971	33.0	67.0		100.0
				forest and other land				
1930	7,495	120,955		128,450	5.8	94.2		100.0
1940	28,928	115,146		144,074	20.1	79.9		100.0
1950	35,958	113,686		149,644	24.0	76.0		100.0
1960	39,800	132,861		172,661	23.1	76.9		100.0
1970	59,383	44,970		104,353	56.9	43.1		100.0

SOURCE: Statistical Office, decennial farm censuses; with adjustments to arable and (in 1970) to grassland as explained in the Appendix.

This strongly suggests that the supply of good land that could be expropriated was becoming scarce, and indeed not a few groups of hopeful ejidatarios discovered that they had been given land which was mere desert.

At the time of the 1970 census, the ejidos and the communities between them owned 49.9 percent of the agricultural area of the country (Table 7.3). As to the particular categories of land, they possessed 61.2 percent of the arable, 49.1 percent of the irrigated area, 37.5 percent of the pasture (counting cultivated grassland as pasture), 67.5 percent of the forest and unproductive land. Thus, in relation to their share in the total area, they did well in respect

TABLE 7.4

EJIDO AND COMMUNITY LAND USE BY REGION: 1970

(in thousand acres)

	Arable	Grassland	Other	Total
Northwest:				
Ejidal	3,430	10,914	9,113	23,457
Communal	312	2,472	1,459	4,243
North:				
Ejidal	2,918	24,300	12,409	39,627
Communal	63	1,600	870	2,533
Northeast:				
Ejidal	1,238	3,771	3,313	8,322
Communal	7	116	86	209
North-center:				
Ejidal	2,434	7,717	5,970	16,121
Communal	72	178	126	376
West-center:				
Ejidal	6,237	8,750	6,056	21,043
Communal	1,064	1,877	3,079	6,020
Center:				
Ejidal	3,353	2,394	2,462	8,209
Communal	158	258	604	1,020
Gulf-south:				
Ejidal	5,402	5,098	6,807	17,307
Communal	635	2,538	5,138	8,311
Peninsula:				
Ejidal	1,245	993	13,253	15,491
Communal	0	0	0	0
Mexico:				
Ejidal	26,257	63,937	59,383	149,577
Communal	2,311	9,039	11,362	22,712

SOURCE: Statistical Office, farm census of 1970.

to cropland but poorly in respect to pastures. It is sometimes alleged that over the years the land redistribution programs have discriminated against the ejidos, donating them mainly the less productive areas. The figures just cited do not support this contention. For the ejidatarios, as small-scale farmers with a predominant interest in basic crops, it is more important to have arable land than grassland. Nevertheless, we cannot be sure that within the arable the ejidatarios have obtained their due share of the high fertility lands.

The regional differences in land use by ejidatarios reflect the differences in types of land available. That is to say that in regions, such as the northern ones, where grassland predominates, it also predominates in the ejidos of those regions; likewise where cropland predominates, as in the center, the ejidos also possess mainly cropland.

In considering these comparisons, we need to remember that the ejido and community land supports a much larger population than does the private farm land. According to the calculation presented earlier, these two groups accounted for 1.8 out of the 2.4 million farm operator families. Thus with 75 percent of the families they possess 50 percent of the agricultural land. However, to advocate a policy of increasing their land holding until it more closely matches their demographic share, would have to take account of the damaging consequences on the volume of food production, as will be seen presently.

This problem of disadvantageous man/land ratios can also be looked at from another angle, namely that of the sizes of the farm units. While in the U.S.A. it is customary to classify farms according to the volume of farm sales, this is not practicable in Mexico, not because of any lack of figures of farm sales, but because these figures are too untrustworthy to be used, especially in regard to their grievous underestimation of the quantity and value of livestock products. We are therefore thrown back on classifying farms according to their number of acres, distinguishing where possible the irrigated from the rain-fed crop farms, and the grazing ranches from both of these.

Dealing first with the private sector, from census to census the statistics record the progressive break-up of the large estates, until in 1970 there remained less than 6,000 farms with more than 500 acres of cropland each. At the same date it is possible to calculate that there were just over 17,000 ranches having more than 1,250 acres of grassland each, and these averaged 6,670 acres in size. To some readers a unit of 6,670 acres may sound large, but in the circumstances of Mexico this is not the case. The ranching areas of the north are so arid and the grazing is of such low quality, that a ranch of this size might support no more than one hundred head of cattle.

Not only the large estates but also the mini-farms have been diminishing in number. Between 1940 and 1970 the proportion of part-time and mini-farms (those under 12.5 acres) to total farms in the private sector declined from 87 percent to 68.6 percent while the area they occupied fell from 2.8 million to 2.3 million acres (Table 7.5). Although this result may have been partially influenced by a more rigorous exclusion in 1970 of units not genuinely agricultural, it also reflects a real trend. Two things seem to have happened. A considerable number of small operators decided to sell their farms, in most cases to other farmers, though those fortunate enough to have near-city locations could sell for building lots. The others, while renouncing farming as an activity, preferred to retain the ownership of their land but rented it for cultivation by neighboring farmers so that census-wise these units disappeared in mergers.

Let us now look at the situation among the ejidatarios. Sixty years ago, leaving aside the comuneros and those who wanted to continue as hired workers, there were probably some 1.5 million families desirous of becoming ejidatarios. After sixty years of redistribution the ejido sector has obtained more than 150 million acres, i.e., some 100 acres for each one of the original families, which would have been quite sufficient land to provide any hardworking family with an acceptable standard of living. But the demographic explosion intervened: in the newer ejidos the parcelas handed out have been notably smaller than in old established ones; while in these latter the ejidatarios were dividing, openly or occultly, their parcelas among their numerous sons.

TABLE 7.5

SIZE DISTRIBUTION OF FARMS BY TYPE OF TENURE

(according to amount of arable land held)

Type of Farmer	Part-time (to 2.5 acres)	Mini-farms (2.5 to 12.5 acres)	Family Farms			Larger Farms (over 250 acres)	Total
			small (12.5 to 25 acres)	medium (25 to 62.5 acres)	large (62.5 to 250 acres)		
			Number of Farms in 1970 (thousands)				
Ejidatarios	280.5	792.8	484.0	144.6	16.7		1,718.6
Comuneros	46.7	64.1	13.6	3.3	1.3		129.0
Private farmers	281.1	286.0	99.9	80.3	60.6	17.0	825.0
Total	608.3	1,142.9	597.5	228.2	78.6	17.0	2,672.6
			Area in Farms in 1970 (thousand acres)				
Ejidatarios	473.2	6,146.0	9,199.5	5,396.6	1,543.1		22,758.4
Comuneros	79.1	435.2	255.2	145.5	143.6		1,058.6
Private farmers	410.4	1,947.6	1,894.5	3,310.4	7,702.7	10,397.2	25,662.8
Total	962.7	8,528.8	11,349.2	8,825.5	9,389.4	10,397.2	49,479.8
			Percentage Distribution of Number of Farms				
Ejidatarios:							
1940	9.1	34.9	43.7	9.8	2.5		100.0
1970	16.3	39.1	35.2	8.4	1.0		100.0
Private:							
1940	42.8	44.2	6.2	5.0	3.1	0.7	100.0
1970	34.0	34.6	12.1	9.7	7.4	2.2	100.0
Total:							
1940	25.6	38.5	25.3	8.7	1.6	0.3	100.0
1970	22.1	42.4	22.9	8.8	3.1	0.7	100.0
			Percentage Distribution of Area in Farms				
Ejidatarios:							
1940	1.0	14.9	47.8	22.4	13.8		100.0
1970	2.1	20.6	46.8	23.7	6.8		100.0
Private:							
1940	2.1	14.9	7.0	12.1	22.3	41.6	100.0
1970	1.5	6.9	6.7	11.7	27.3	45.9	100.0
Total:							
1940	1.6	14.9	26.2	17.0	18.3	22.0	100.0
1970	1.7	13.0	24.6	17.1	18.1	25.4	100.0

SOURCE: Statistical Office, farm censuses of 1940 and 1970.

NOTE: Collectivized ejidos and communities are excluded.

In the "percentage distribution" sections, the comuneros are included with the private farmers as whole communities (all of them over 250 acres in size).

In the "percentage distribution" sections, the ejidatario size groups are: 0 to 2.5, 2.5 to 10, 10 to 25, 25 to 50, and over 50 acres; whereas the private farm sizes are: 0 to 2.5, 2.5 to 12.5, 12.5 to 25, 25 to 62.5, 62.5 to 250, and over 250 acres.

At the end of the Cárdenas era in 1940 some 56 percent of all parcelas exceeded 10 acres in size and accounted for 84 percent of all ejido arable land. Of course some regions fared better than others: the northwest and the north already at that time had larger than average parcelas, whereas in the states of the center they were small.

In the year 1970 parcelas over 10 acres represented only 44.6 percent of the total and only 77.3 percent of the area. Two of the most fragmented states were Hidalgo and México. In the latter only 7.7 percent of the parcelas exceeded 10 acres in size and only 30.9 percent of the ejido land was in this group. In Hidalgo the position was almost as serious. However, in the state of México a high proportion of the ejidatarios have full-time or part-time work in industry or commerce, whereas in Hidalgo very few have off-farm occupations so that the poverty implied by this decimation of the parcelas is a human tragedy.

Among the comuneros the fragmentation is still worse. Only 20 percent of all their parcelas are larger than 10 acres. Nevertheless, in this group the prevalent custom is to use the parcela for growing food for the family, rather than for generating income; most comuneros earn their living in the forests are resin-gatherers or lumber-workers.

From this analysis of farm sizes two conclusions may be drawn. First of all, while the number of mini-units has been declining in the private sector, it has been increasing in the ejidos and the communities. Secondly, in the private sector, although a steady elimination of big farms (over 500 acres) has been going on, the importance of those in the 50 to 250 acre group has become pronounced, as a result of purchases and mergers. By contrast, in the ejido/community sector, where purchases and mergers are prohibited, we find hardly any farms in this size group.

It can be argued from a social point of view that the policy pursued with respect to the ejidos has had beneficial effects. By permitting or condoning fragmentation it enabled the next generation (the sons of the original ejidatarios) to remain on the land, thus delaying their emigration to the cities where they might or might not have found employment. It satisfied the land hunger but was negative in respect to land use. Multiplying the number of land users who lack both education and economic resources has blocked the way to improvements in land productivity. If, on the other hand, the ejidatarios had been allowed to build up units of 50 to 250 acres, as in the private sector, land productivity would have risen and so would the incomes of the ejidatarios (though there would have been fewer of them).

From the national point of view, the provision of production capital, in the form of land, to the marginal members of society can be justified when the population is small and land is still plentiful. When these circumstances change, as they have done in the Mexico of today, the policy merits reconsideration. Land has become one of the scarcest resources and its optimum use has consequently become a priority.

We may conclude this section on distribution and fragmentation by guessing briefly at the likely trends in the coming years. Since 1970 there has been some further creation of ejidos, notably in the northwest and the southeast, but not on a scale in any way comparable to that of the previous decade. The opportunities for the eighties are equally meager if not more so. Further areas may be opened in the peninsula, and a certain number of law-evading landowners may be caught, but neither of these sources would produce substantial quantities of land for distribution. A new agricultural plan, the *Sistema Alimentario Mexicano* (SAM), launched by President López Portillo in 1980, provided that those parts of cattle ranches deemed by Department of Agriculture officials to be suitable for crop production, might be expropriated and donated to ejidos. At this point it is not known to what extent the authorities will make use of these powers. Otherwise, the only significant new land supply for ejidos would have to come from changing the law and lowering the permitted upper limits of farm size. Such a policy has indeed been advocated in radical circles, but the damage to food production would be so great that it seems unlikely that a government would resort to this measure. Aside from that eventuality, it may be assumed that the stratification of farm operators according to their tenure status will remain in the coming years much as it is at present.

On the other hand, within each tenure group the recent trends in respect to size may be expected to continue. In the private sector mini-farms will continue to disappear as more and more of their operators find alternative employment. The farms in the 50 to 125 acre category are likely to increase in number as a result of mergers; but perhaps there will be few takers for those near the maximum legal limits because of their vulnerability to invasions and expropriation.

Among the ejidatarios and comuneros, large families remain the rule, and the numerous sons have to be accommodated somehow. Inevitably fragmentation will continue until such time as the natural increase is balanced by outward migration. Mergers of parcelas in ejidos or in communities will continue to be forbidden, thus preventing larger units from being formed. Enterprising and skilled ejido farmers have to move into the private sector.

CONTRIBUTIONS TO PRODUCTION

Having described Mexico's farm population, its modes of tenure and the amounts of land at its disposal, we can now try to discover what contribution to national agricultural production is being made by each principal group of operators. This topic may be examined first in regard to total output and then in relation to certain key commodities. But as a preliminary step something needs to be said about subsistence farming which is still practiced in Mexico on a significant scale.

The evidence on this subject is unfortunately limited and far from reliable. However, the 1970 census reported the number of production units having sales of less than 1,000 pesos (80 dollars) in 1969. Of these units 799,000 belonged to ejidatarios or comuneros and 646,000 to private farmers. But this tabulation of sales referred exclusively to crop products, and we have good reason to suppose that a considerable proportion of these apparently subsistence crop farmers were in fact commercial sellers of meat or milk or eggs. Evidence as to the number of persons engaged in poultry keeping and hog rearing indicates that we should deduct about half a million on this score. Furthermore, in the second section of this chapter we calculated that some 500,000 mini-farmers obtained more than half their income from nonagricultural sources. There might be some overlapping between this group and the one just mentioned, but deducting this second half million (or slightly less) leaves us with 500,000 real subsistence farmers, defined as persons who sell nothing, or very little.

Some additional evidence exists in the family consumption survey carried out in 1968 under the auspices of the Bank of Mexico. Although the published material does not reveal what proportion of the food was purchased and what proportion home-produced, one can make certain common sense assumptions for particular commodities, especially corn, beans, vegetables, fruit, milk, and eggs. Using this approach, we find that the lowest income group of the survey registered approximately 80 percent self-sufficiency, measured in value terms. In the next highest income group the percentage was already much lower.

According to the survey's estimates, only 10 percent of the farm population belonged to this lowest income group and could conceivably be classified as subsistence farmers. By taking the other approach, we arrived at around 500,000 families, representing 15 percent of the farm population. In the absence of more reliable information, it seems reasonable to combine these two pieces of evidence and conclude that something between 10 and 15 percent of the farm families produce mainly for their own needs, and have only a weak and intermittent relationship to the market.

Let us now pass on to examine how the total agricultural output is divided among the different farm groups. In 1969 according to the census* the private sector produced 56.5 percent of the total output of agriculture and forestry, the ejidos 41.6 percent and the communities 1.9 percent. To these figures an important reservation has to be attached, namely that the census underestimated, probably by as much as half, the volume and value of the livestock products produced. Since two-thirds of the herds and more than two-thirds of their output belong to the private farm sector, a correction for this census error would appreciably raise the private farmers' share and reduce that of the ejidos. Their true shares were probably close to 60 and 38 percent respectively.

The census also tells us something of how the output was distributed among farms of different sizes. For instance, of the total output of the private sector only 16 percent came from the mini-farms of less than 12.5 acres; whereas in the ejido/community sector 35 percent of their output came from under 12.5 acre parcels (excluding collectivized ejidos and communities from the calculation). Once again this shows the greater fragmentation within the ejidos and the absence of any units in the large size groups.

---

*An adjustment had to be made to distribute among the groups the livestock products which the census recorded as emanating from built-up areas (see Appendix).

Perhaps even more important from the viewpoint of efficient resource use are the differences between the groups in respect of productivity. If we relate total production to the total area of arable land (but subtracting that part of the livestock production calculated to have been contributed by grassland), then we find the average output per acre to be twice as high in the private farm sector as in the ejidos and communities. Moreover, the historical trend has been for this gap to widen. Between 1950 and 1970 average output per acre increased more than threefold in real terms in the private farms but by less than 60 percent in the ejidos. No one factor explains this startling difference; it has been a combination of technical improvements, the adoption of new and more profitable crops and greater attention paid to quality and marketing.

An even greater contrast emerges when we measure productivity in terms of output per person. In this respect the private operators' output was on average four times that of the ejidatarios and comuneros in 1970, and quite certainly the ratio for the ejidatarios was no better in 1980. Once again the gap has been widening during recent decades, though not so rapidly as in the case of output per acre. The explanation includes the factors mentioned above, but in addition the greater degree of mechanization on private crop farms, plus the greater importance among private farm activities of ranching with its low manpower input.

One further comment should be made. The acquisition by the ejidos during recent decades of large quantities of grassland ought to have enabled them to achieve a significant increase in livestock numbers and in meat and milk production. However, this did not occur in the proportions that might have been expected. The private farms of over 12.5 acres increased their cattle numbers at almost the same speed as the ejidos, indicating an intensification of pasture use in the private sector, whereas the ejidos had fewer cattle per 100 acres of pasture in 1970 than in 1950. This suggests that the communal management of pastures within the ejido system has not, on average, led to major improvements in their quality and productivity. As a consequence the incomes of ejidatarios have not increased as much as they should have done given the increase in land resources at their disposal.

FARMERS' INCOMES

A central issue among the problems of the agricultural sector is the level of farm incomes or, more accurately, how the farm population is distributed between low, medium, and high income groups. Unfortunately, in Mexico there is no systematic information on this topic. All that can be done is to piece together evidence, none of it wholly reliable, from a number of sources and evaluate the resulting picture.

One source is the publication in the census of tables showing the value of production and the operating expenses, so that the difference between the two should represent "net income" using this phrase to cover the reward for the operator's own work and that of his family, depreciation on farm assets and interest on own investment in the farm. The 1970 average net income in each group was:

Private farms over 12.5 acres	19,460 pesos
Private farms under 12.5 acres	918 pesos
Ejidatarios per parcela	4,426 pesos
Comuneros per parcela	2,682 pesos

The net income per ejidatario surpassed considerably that of the average comunero, partly because the comuneros' parcelas are smaller and generally their land is of poorer quality. The private mini-farms registered low net incomes mainly because the group includes so many land occupiers who are not really farmers. As to regional differences, the highest incomes were recorded in the northwestern states and the lowest in those of the center, this being the case in each and all of the operator groups.

Greater detail is available for ejidatarios and comuneros according to the size of their parcelas. For instance, here are data for ejidatarios:

Acres	Pesos
Up to 2.5	875
2.5-12.5	3,129
12.5-25	6,272
25-50	9,881
50	24,952

In each size group except the highest the incomes of the comuneros were lower. Since at that date it would have been impossible, even for a poor family, to live on less than 1,000 pesos per year, it seems fair to conclude that operators with very small parcelas depended chiefly on nonfarm sources of income.

The second source of information is the survey of family consumption sponsored in 1968 by the Bank of Mexico. The survey showed that 10 percent of farm families had incomes of less than 3,600 pesos per year, whereas the census indicated that 40 percent of all production units had net incomes of less than 2,300 pesos. The difference is considerable but explicable. First, as noted earlier, the census badly underestimated the output of livestock products which affects more especially the small farms. Secondly, the census was recording only income from farming operations while the consumption survey reported income from all sources, and indeed mentioned that in respect of farm families only 45 percent of their incomes were derived from farming in the poorest groups. Perhaps the worst situated were those whose land holdings were small and who nonetheless could not obtain supplementary employment.

So far the attention has been focused on farm operators, but indications are available with regard to hired farm workers in the data published by the National Commission for Minimum Wages. A different minimum wage (one for farms and another for other occupations) is set for each of formerly 105 districts (reduced to 89 in 1980), the differences purporting to reflect differences in the cost of living. Thus, the highest figures are found in districts adjacent to the U.S. border and those around the largest cities, especially Mexico City. The lowest minima are in certain districts of the south— Guerrero, Oaxaca, and Chiapas.

In 1980 the arithmetical average of all the districts was a wage of 125.80 pesos per day, equivalent to just over 5.50 dollars. But the level in the high-wage districts was nearly double that prevailing in the poor remote areas. Although these wage rates may appear low, they in fact represented a substantial improvement over previous epochs. Thus, in the twenty years from 1955 to 1975 the agricultural minimum wage had actually doubled in real terms, a quite remarkable achievement; it increased a further 50 percent in real terms between 1975 and 1980.

It can of course be objected that this information does not offer a fair indication of what farm laborers earn, because in many districts what they are actually paid may be considerably less than the statutory minimum. While this is undoubtedly true, it is also true that in the more modernized districts, many of the progressive farmers pay wages in excess of the minimum in order, when they have acquired skilled and conscientious workers, to retain their services year after year. Thus while the statutory minimum wages probably exaggerate the incomes of hired workers in some of the more backward areas, they probably underestimate the prevailing wage in the advanced areas.

A comparison of the trends in agricultural minimum wages and those in the minima for nonagricultural occupations is also informative. During the twenty years to 1975, while the agricultural wage doubled in real terms the nonagricultural rose from 100 to 251. This is less of a difference than many readers might have expected, in the light of what was happening in other countries. It suggests that the Mexican authorities were trying to prevent the opening up of too large a wage gap. But to what extent they were successful in practice it is impossible to say. There are some who assert that the minimum wage legislation, under Mexican conditions, merely impeded the hiring of persons whose productivity could not match the wage stipulated, and consequently aggravated rural unemployment.

One also has to remember that the significance of a "wage per day" depends on the number of days per year that the worker is employed. Except in irrigation districts and some tropical plantations, most agricultural employment is seasonal, and this is especially the case in zones where farming remains primitive and the wage minima are low. Some of the migrant workers may find employment during only 20 or at most 30 days during the year, in which case a wage packet of say 125 pesos per day (1980) was very modest. But the great majority of these people are members of farm operators' families—ejidatarios and others—and what they earn seasonally they regard as supplementary income to that derived from their farm.

During the past two or three decades great advances have undoubtedly been achieved in the incomes of farmers and farm workers. Also, without doubt, the advance has been unevenly spread. From a situation in which poverty prevailed in almost all parts of the country, certain groups have moved forward appreciably, partly because farm operations were themselves being modernized and partly because with urbanization alternative job opportunities were multiplying. The geographically more remote regions, notably in Guerrero and Oaxaca, have participated very little as yet in this forward march, and the incomes of their inhabitants remain low. We cannot quantify either the depth or the extent of this poverty, but some qualitative indicators are cited in the following section.

LEVELS OF LIVING

The difference between levels of income and levels of living can be identified in one sense as the difference between the quantity and the quality of satisfactions. Of course, the two are intimately related. A person whose income is very low cannot afford to purchase for his family an adequately balanced diet, cannot acquire a house with paved floor, piped water and drainage. However, other satisfactions depend mainly on services supplied by the public authorities: schools, clinics, electricity, municipal water and drainage, radio and TV network, civic security and so on.

It has long been realized that levels of living cannot adequately be measured by the yardstick of per capita national income, even when inequalities of income distribution may have been taken into account, because income by itself is only one of the components of well-being. One could imagine a situation in which for a period of years no progress occurred in the amount of purchasing power in people's pockets, but very considerable improvements in education and other social services. In the following paragraphs both aspects will be commented on; first a brief glance at patterns of food consumption and then an account of the incidence of the more social qualities of living.

In the previously mentioned survey of families, the distribution of consumers' expenditure, for example, follows the normal pattern, that is to say that food takes two-thirds of family expenditure in the lowest income groups, falling to less than 25 percent in the highest income group, and this is true equally of agricultural and nonagricultural families. The only notable difference between the two sectors was that in the lowest income group the agricultural families spent less on imputed rent and more on clothing than the nonagricultural.

When the consumption of individual foodstuffs is examined, families in the lowest income group have a diet composed almost exclusively of corn and beans, supplemented by very small amounts of other commodities. However, in this income group the nonagricultural families consume rather more fruit, fats, and meat, partly probably because of availability in urban stores and partly because dietary patterns begin to change in urban surroundings with a higher proportion of sedentary employment.

When one examines the 3,000–6,000 peso income group (of 1968), comprising only 3.5 percent of farm families but 19 percent of nonfarm ones, the differences become more striking. Farm families at this level of income, and they are among the richest farmers, consume less corn and beans but more per capita of everything else compared with the poorest families. The differences are especially significant in wheat, fats, and milk. Nevertheless they continue to consume more corn and beans than nonfarm families of the same income group. This confirms the view that dietary habits change much more slowly than financial circumstances.

It is particularly striking that among farm families of all income groups the consumption of vegetables, fruit, and livestock products is relatively low, although these are commodities which can easily be produced on farms. In this the Mexican farmer is quite different from his Asian counterpart. An effort is now being made to modify this nutritionally disadvantageous situation through a government program for family gardens, promoted by female extension workers; but the traditional hostility to gardening will not be quickly overcome.

Other indicators of welfare have been recorded and tabulated by the National Commission for Minimum Wages. Among these we have selected the following eight which appear to manifest low levels of living:

1. more than 80 percent of economically active population occupied in the primary sector;
2. minimum farm wage less than 24 pesos daily in 1975;
3. literacy rate below 55 percent;
4. primary school attendance below 50 percent;
5. less than 15 percent of dwellings have inside water supply;
6. less than 15 percent of dwellings have piped drainage;
7. less than 25 percent of dwellings have electricity; and
8. less than 40 percent of families have radio.

No less than 15 of the 105 districts registered poverty according to one or more of these indicators; 10 districts appeared in four out of the eight lists. Not surprisingly the worst placed districts were almost all located in a block of southern mountainous country stretching from Guerrero into Oaxaca. Slightly less deficient were two districts in Chiapas and one each in the state of México, Querétaro, and the south of Nuevo León.

One further comment on rural poverty. States which today are the most disadvantaged were also among the ten least favored some twenty years ago when the author compiled a similar welfare index. But what is more interesting is that several states like Zacatecas, Tlaxcala, Hidalgo, and Tabasco, which then registered extreme backwardness, have totally disappeared from the list. Undoubtedly a process of catching up has been taking place, some of the poorer states having made great forward strides. This process still persists and it will be aided by the present government's program for the creation of new urban centers, by progress in the construction of rural roads, and finally by outward migration from the most marginal regions.

So far we have been considering levels of income, of social services, and of cultural attainment. But there are other aspects which enter into the sum total of satisfactions though they do not lend themselves to being quantified. Among these nonmaterial components probably the most significant are the human relationships prevailing in rural areas. A community can be harmonious if its goals are not conflictive and if its members communicate easily and frequently with one another. We noted earlier that in Mexico one of the unintended consequences of the Agrarian Reform has been to create, over the years, a deep cleavage among the several groups of farm operators. This cleavage, which accentuates social barriers and at times erupts into violent conflicts, is a negative component in the level of living index; if the index is to be improved, one of the necessary actions will have to be to reduce these barriers and conflicts. The old saying that "a nation divided against itself cannot stand" is as relevant to the Mexican farm sector as it is to society as a whole.

The social atmosphere in rural communities also has a direct impact on human motivation for improvement. In previous chapters much has been said about the supply of land, the potentials for improving its productivity, the integration of crop and livestock activities, the dissemination of technical know-how, the provision of credit, and so on. But in all these matters the motivation of each individual farm operator will be a vital, and in many instances the decisive, factor. Whether the farmers are willing to save and/or borrow in order to invest in incrementing the productivity of their physical and economic assets will depend on the extent to which the system provides opportunities for the advancement of themselves and their families. This also determines the amount of hard work and personal dedication they apply to the management of their farms.

At the moment in Mexico two systems are operating side by side, one having at least ostensibly the usual incentives of a market economy and the other combining vexatious obstacles to personal advancement with powerful disincentives to technical improvement. The following chapter then is devoted to a consideration of how the juridical and institutional framework might be relieved of some of its more troublesome rigidities in order to ameliorate the problem.

8. A New Agrarian Reform

The question has to be asked: can the nation afford to have an agricultural sector in which the system of incentives to efficient farming has ceased to function? The private sector and the ejidal sector, each for different reasons, are stagnating. In neither is there the human mobility which creates progress. Private farmers are under constant attack in the left-wing press; the authorities too often are unwilling to distinguish between the legitimate and the illegitimate operators. On the other hand, ejidatarios suffer simultaneously from discrimination and from overprotection. They cannot expand their parcelas, and they cannot obtain credit (except by forming societies). Their subsidies and tax exemptions absolve the less than mediocre from having to improve themselves. It is imperative to face frankly these dilemmas and try to formulate suggestions for giving more scope to the capacities which many of the farm people undoubtedly possess.

This is no new problem. Its existence has been recognized for many years and various attempts have been made to deal with it. During the regime of President Cárdenas (1934–40) a concerted effort was made to organize some of the newly created ejidos in the form of Russian-style collectives. Some of these prospered for a few months, others kept going for several years, but almost all of them finally reverted to individual family producing units. Again during the term of President Echeverría (1970–76), a considerable body of opinion in governmental circles advocated the collectivist solution. Those who took this view despaired of being able to impart a sufficient amount of know-how to two million ejidatarios; they saw the mini-farm as a barrier to progress which must be abolished—the land must be put under competent technical management. Under the administration of President López Portillo this philosophy has receded to the universities whence it came. However, there are many who expect it to stage a comeback at some future date; hence some attention has to be paid to the pros and cons of such a program in the Mexican setting.

The collectivization of an ejido can take various forms. In some instances only a small group of ejidatarios agree to work together, in others a large group, and in yet others the entire membership of the ejido. Their agreement to operate collectively may extend to certain crops and not others, or to certain classes of animals only, or it may cover all the ejido's activities. In travelling round the country one may encounter collectivized crop cultivation on only 100 acres or on 25,000. One may find collective cattle herds of 50 head or as many as 5,000. For administrative convenience the authorities have preferred trying to collectivize entire ejidos.

Of the several hundred collectives created during the Echeverría administration, many were still operating at the close of the seventies, so that their successes and difficulties could be readily observed. Let us begin with the technical aspects of the system, leaving the human ones for consideration later.

The first and perhaps most emphasized advantage of collective production is the pooling of the parcelas. The old boundaries between the ejidatarios' strips of land are obliterated in order to create compact fields of 25 up to 500 acres or more in size, a transformation which facilitates the adoption of modern methods of cultivation.

The second advantage is the opportunity for using improved seeds, adequate quantities of fertilizer and pesticides, correct cultivation practices such as depth of ploughing, date of sowing, etc. Similarly the organization of cattle or hogs in large collective units facilitates the adoption of correct feeding practices, improved stabling, and more effective veterinary control.

A third benefit of operating large crop units is that it permits the replacement of work animals by tractors, combines, and other machinery. This on the one hand ensures the better performance of the various operations of cultivation, and on the other liberates for cash crops a considerable number of acres which previously had to be dedicated to growing food for the work animals. Furthermore, a large ejido can establish a workshop for servicing its machinery and undertaking repairs.

A fourth advantage which large production units enjoy is the opportunity for purchasing in bulk the materials they need—seeds, fertilizers, feedstuffs, etc.—and for selling in bulk the products of the farm. In both cases the prices obtained will be more advantageous than those offered to the small operators. In truck crops, large-scale production makes on-farm grading and packaging an economically viable activity.

Fifthly, the large unit can be a more attractive client for bank credit. It is likely to have a technically qualified administrator, it keeps accounts, its assets and liabilities can be readily identified. Being more creditworthy than the individual ejidatario, the collective can build up the productivity of its capital through land and livestock improvement.

Another advantage is that the collective ejido should be able to offer more employment than was generated by individual operations. This becomes possible partly through being able to grow a greater variety of crops, partly through having a larger number of animals, and partly through supplementary activities of grading, packaging, and processing where these are undertaken.

Finally, the collective should ensure a far higher level of technical management, since it has the services of trained agronomists and can call in specialists for advice on special problems. Superior management in turn leads to more economical operations, higher quality in the commodities produced, better prices for the produce sold, and thus improved incomes for the unit and its members.

All this adds up to an impressive list of benefits. Indeed, if this were the entire story and a faithful picture of what occurs on collectives, the system would have spread throughout the country like a prairie fire, and the members of the collectives would be the happiest and most prosperous producers in the Republic. The fact that many collective ejidos are not so prosperous, that many face serious difficulties, and that not a few have collapsed, suggests that there is another side to the picture which must be examined.

First, it has been the common experience all over the world that large agricultural units are more vulnerable than small ones to the hazards of nature. When a fungus disease, for example, enters a small plot of rice, it may not spread to other plots if they are some distance away; but when it enters a unit of several thousand acres it can cause immense damage. Similarly with livestock: if an ejidatario has two cows infected with mastitis, it is not inevitable that his neighbor's cows will be infected. Whereas, when several hundred cows are stabled together, any highly infectious disease will quickly spread through the entire herd. Unfortunately, the incidence of disease is high in collective livestock ejidos.

In the second place, any error of judgment by the manager of a large unit may have dire consequences. If he decides to plough too deep, chooses a variety of seed which germinates badly in that particular locality, fails to apply the pesticides in good time, omits to note the first symptoms of disease among the hogs or the cows, the ensuing losses may be on a dramatic scale. Mismanagement by an individual ejidatario affects only his small unit. Since even the most skilled and experienced farmers inevitably make wrong decisions from time to time, and since in Mexico these skills are scarce, the chances are rather high that such errors will occur on collectives.

Thirdly, collective farms are almost invariably overmechanized. The quantity of machinery for a new collective is decided, not by the amount that the farm can afford to spend, but by an enthusiastic young agronomist who consults his textbook as to the optimum tractor park. He is spending not his own money but that of the credit bank which has received instructions to support collectives and not to worry about eventual bad debts. And it is always better to be overprovided, so that if some machine breaks down at a critical moment, another can be brought into use.

The introduction of machinery naturally reduces manpower requirements. This would be advantageous in circumstances where manpower was scarce and where alternative employment was readily available. In Mexico neither circumstance exists. For instance, when a typical ejido of 2,500 acres and 300 families becomes collectivized and mechanized, what happens to those families? At first the problem does not present itself, because there are new houses to be built, land to be cleared, stables and workshops to be constructed, farm roads to be made. But when these initial tasks have been terminated, the ejido manager can find nothing to occupy perhaps half of these families. They have nowhere to go, they have a right to stay, and they have to be paid for doing nothing, while the farm has to bear large expenditures for machinery maintenance and amortization.

Another drawback of large collectivized units is their dependence on the reliable functioning of external services. When in mid-harvest a combine harvester breaks down, it probably cannot be repaired by the ejido or by any mechanic in the nearby village or small town, whereas the ejidatario carries on with his ox or mule. Large units also depend on prompt and reliable transportation services. Collective farms in Mexico are not equipped with capacious silos but expect the arrival of trucks which may or may not materialize during the peak days of harvest. The dairy herd depends upon the daily appearance of the refrigerated milk tanker—if it does not come, the milk sours. But the small operator relies on no one but himself and his donkey to transport produce to market. In developing countries, these external services tend to be more than capricious.

For the successful management of a large agricultural enterprise, it is necessary to combine in a single person a wide range of capabilities. In Mexico one finds collective ejidos of 50,000 acres or more being operated by young agronomists only recently out of college and without experience of the soils or climate of the district to which they have been assigned. They come face to face with maladies of crops or animals with which they are unfamiliar. The specialist is far away and otherwise engaged. Meantime the losses can be alarming.

Moreover, the manager faces problems for which he has not been trained. He has to deal with the credit bank of which he is generally an employee; he has to tolerate the interference of the officers of the ejido's assembly who may be the local political bosses; he has to bow to directives, often ill-conceived, from federal government departments. Caught in such crossfire, it is surprising that so many of them maintain their cheerfulness and devotion to responsibilities.

This review of the salient technical advantages and disadvantages of collectivist production might incline those Mexicans who still advocate this course to reconsider their positions. Whatever might be the theoretical benefits, in the practical circumstances of a country like Mexico the disadvantages outweigh the gains. From the evidence available on ejidos in different parts of the country, it is an illusion to suppose that to collectivize is the royal road to rapid and successful technification. Nor is this finding in any way surprising, since it repeats the experience which has been learned the hard way in all those countries which have tried this formula.

Moreover, technical know-how is only part of what goes to create enterprise efficiency. Equally important is the motivation of human beings to display initiative, to be interested in innovation, to take risks, to combine input factors in ways which maximize profit and personal reward. The collective organization of production pays little heed to human motivation and prefers discipline to initiative.

Consider first the proposition that collectivization should be a voluntary transformation, to be undertaken only when it is desired by the majority of the members of an ejido. In real life situations it is difficult to determine how much freedom of choice individual ejidatarios really possess. It is rather easy for educated technicians and/or strong-armed local bosses to lean on ill-informed and mostly illiterate operators of parcelas and to paint a brightly colored picture of an escape from individualist misery to collectivist prosperity.

In persuading the ejidatarios to adopt the new system, it is emphasized that the collectives will be *their* enterprise, which *they* will operate, and which will develop according to *their* decisions. They are not told that the role of the assembly will be to rubber-stamp the decisions taken by the manager and/or the political leaders. It would indeed be disastrous to place real decision making in the hands of a group of untrained and uninformed ejidatarios. The direction of these large collectives must be authoritarian if they are to stand any chance of success. The members of the ejido soon realize that the so-called collective really belongs to some impersonal authority for which they are expected to labor.

In most cases a collective ejido pays the same wage to each of its members, albeit accepting some differentiation for skills such as tractor driving. Indeed they are paid whether there is any work for them to do or not, and whether the ejido has any income or not, as has happened in some of the ejidos of the southeast. This system, however, provokes discontent. Those who work hard perceive that their efforts merely benefit the lazy; hence either they try to have the system changed or they cease to work hard.

Where collectivization is applied to an already existing ejido, the distribution of income usually reflects, at least partially, the quantity of land or animals which each member has contributed, i.e., the member who has brought 20 acres receives a higher wage than the one who has brought only two. After a time these distinctions tend to be forgotten, and the principle may be changed into one of equal shares for all. (In some quarters it has been argued that the system of payment on the basis of original land contribution should be retained indefinitely, because this would enable those to whom the ejido can offer no work to obtain employment outside. What a strange suggestion! Here are collectivizers proposing to create a class of absentee landlords, collecting rent from the collective for its use of their former parcelas—parcelas which they never owned but merely held in usufruct.)

At the end of the seventies a new formula was elaborated as a possible solution to the problem of fragmented ejidos. It was entitled "cooperative association between the private sector and ejidatarios," the idea being that a private investor on the one hand and a group of ejidatarios on the other should form a cooperative for cultivating the ejidatarios' land which would be pooled into a single unit. The ejidatarios who volunteered could be merely a dozen or twenty or more, though their parcelas would have to be more or less contiguous; but rarely would all the members of an ejido join if the reality of *voluntary* participation were strictly observed. The private sector element could be either one or more private farmers, a corporation engaged in food processing or distribution or some other investment group.

In some respects the arrangement would resemble that which in other countries is known as "vertical integration" where a food processor contracts with a group of farmers to provide the raw material for his processing plant: e.g., beans for freezing, broilers for supermarkets. In other respects it would resemble a small-scale collective, with the important difference that the management would be in the hands of profit-motivated persons having every incentive to employ the best available technology. Such units, therefore, should stand a good chance of surpassing in efficiency the collectives hitherto run by bureaucrats.

On the other hand, it is not clear how the labor aspects of the cooperative would be handled. While a private farmer can fire a lazy or incompetent laborer, a cooperative cannot dismiss one of its members; and if the hard-working members perceive one or two expending much less effort but still drawing their shares of the profits, the morale of the cooperative sinks and, if tensions increase, the cooperative may disintegrate.

This brings us to the crucial matter of human relationships in collectives. In considering this it is pertinent to go back in our minds to the epoch of the Revolution and remember that one of the institutions which the revolutionaries sought to overthrow was the social nexus between the hacendado and his peons, the latter being economically bound to their employers. Strangely enough, the collective ejido recreates precisely the same relationship: on the one side the management whose directives have to be obeyed and on the other the peons (retitled "members of collective") who cannot quit without losing all they have—their land, their homes, and their jobs.

In certain instances where new ejidos have been created, some concessions have been made to the ejidatarios' instinct to possess property, and the members have been allotted five acres each for their private use. This occurred, for instance, in the Chontalpa (Tabasco) project where, while the livestock were operated collectively, the members were given private plots on which to cultivate corn and other crops. This exactly reversed the system adopted in the communist countries where, while private plots up to five acres are allotted for keeping animals and growing vegetables, it is recognized that crops are more advantageously cultivated on large units and livestock on small ones. However, in most Mexican collectives it is not the custom to provide private plots at all.

Unimpressed by the Mexican failures and the disheartening experiences in other countries, the protagonists of collectivism will insist that a social form of organization is superior to an individualistic one, and that a socially oriented human being is superior to one who thinks first of himself and his family and who hungers to possess property. But in maintaining this position they should recognize that human beings have been individualistic for thousands of years. Maybe the circumstances of the modern world require that they should be transformed into social animals, but this cannot be accomplished in a decade or two, as Mao Tse-tung admitted. If the attempt is made to introduce a social organization of society before human nature has been changed, it can be accomplished only by the use of force.

PRODUCTION INITIATIVE

In the sense in which the phrase will be used here, "production initiative" means the capacity and the will to make a productive enterprise succeed. It is much more than a combination of skills; it is the drive and the dynamism which are the mainspring of progress. It is also distinct from the concept of management as found in the textbooks, although it features the determination of an individual to secure the optimum utilization of the resources at his disposal.

These resources may be substantial, as in the case of a large production unit, or they may be modest, as is so characteristic of the great majority of Mexican farms. In the latter the operators do not think of themselves as managers, neither are they consciously taking initiatives; they are merely behaving in the manner which to them seems most likely to achieve prosperity. Nonetheless, their actuation does contain the ingredient of production initiative, and the best of them will differ from their neighbors in possessing that capacity in a high degree. It is the difference between an unenterprising traditional farmer and a successful innovating one—a difference which can be observed just as much among mini-parcela ejidatarios as among private farmers on larger properties.

The importance of decision-making on the farm, and of the motivation which inspires it, has been misunderstood and in many quarters underestimated. Yet experience shows that it is the largest single influence in determining the success of an enterprise. Differences in size of farm, in volume of sales, in fertility of soil, in availability of water—all these of course influence the profitability of a farm, but more important than any of them is the human factor. In a situation such as that which confronts Mexico today, when there will be the utmost difficulty in mobilizing sufficient physical resources to feed the nation, the efficient use of these resources becomes a topic of top priority. And, as has been argued, efficient use consists in a mixture of technical know-how and production initiative.

But man is also inherently rather lazy. He has to force himself to make an effort, and he will do this only if he sees attractions in hard work which outweigh its disagreeableness. In short, he needs incentives. If these are strong he will exert himself; if they are weak, he will work only as much as is necessary not to get into trouble. Because he is no altruist, these rewards have to be extremely concrete. For most they have to be in the form of money, a larger income out of which the farmer can obtain for himself and his family some of the objects he has dreamed about.

If he is a self-employed person, as most farmers are, the attainment of these rewards requires two things: skill and initiative in managing the land and other resources at his disposal; and the possibility of expansion, of applying his talents to organizing a larger volume of resources, which is why the more dynamic ejidatarios contrive to add to their parcela-enterprise by purchasing small private farms. In other words, part of the incentive has to be the existence of a ladder up which the successful operator can climb until on the higher rungs he can operate a more rewarding business. Inasmuch as the possibilities of income improvement on mini-farms are rather limited, the more important incentive lies in the existence and efficacy of an agricultural ladder.

There exists a reverse side to this system of incentives, namely a knowledge of the consequences of failure. A self-employed artisan who is inefficient or lazy or both, loses customers until he goes bankrupt. Similarly a private mini-farmer sees his income becoming smaller and smaller until he is forced to sell up and seek other employment. The ejidatario enjoys protected use of his parcela and, provided he continues producing just enough for subsistence, may remain there indefinitely with his family on the poverty line.

Incentives and disincentives are more conspicuously present in some sectors than in others. In agriculture, unfortunately, they tend to be weak in most parts of the third world. There are several reasons for this situation. One is the physical obstacle of land shortage which makes it difficult for an energetic and skillful farmer to expand his scale of operations. Only a small number of farms come onto the market for renting or purchase in any one year, and it will be a lucky coincidence if such a farm becomes available within reasonable distance of the land which our hypothetical farmer is already operating. A cotton spinning mill faces no such constraint. If such a firm had to acquire two thousand acres of land for every 250,000 dollars' worth of sales of yarn, there would not be many large cotton spinners in the world. This is one reason why there are few really large units in farming.

Another problem derives from the hereditary nature of the occupation. Parcelas or private farms pass from father to son, but if the father was a competent farmer, there is no reason for expecting the son to be competent too. Production initiative is not a inherited characteristic. We do not expect the sons of doctors or of architects to follow their fathers' professions, yet it is commonly supposed that anyone can cultivate a piece of land or tend a herd of cows.

Because the great majority of farmers in developing countries are mini-farmers and because many of them are mediocre operators or worse, the result is that a large proportion of the agricultural land in most of these countries is not being farmed competently which, at the present rate of population growth, threatens the world's food supply. If the inefficient cannot be replaced progressively by the more efficient, the prospects are not encouraging for expanding agricultural production at the necessary speed.

How do these general characteristics of the farm sector manifest themselves in Mexico? Let us consider first the private farmers. At the bottom end of this group we can find evidence of mobility in the rapid speed at which the number of very small units has been declining since 1960. Many thousands of part-timers, it is true, are still retaining their land while working elsewhere, but many more thousands have sold out and their land has been merged into larger units. To that extent a process of weeding out is functioning.

In regard to the agricultural ladder, in the days, thirty years ago, when in the northwest and elsewhere land was being opened for cultivation with irrigation, it was possible for energetic farmers and farmers' sons to move from the crowded regions of the country and become colonists or ejidatarios in the new zones. There was a Michoacán peasant who arrived without a single centavo in the Valle de Culiacán (Sinaloa), who worked as a peon until he had saved enough money to buy a few acres, which he built up gradually. Within a few years he had 35 cows, fed mainly on his own alfalfa, with an average daily milk yield of 12 U.S. quarts, 300 hogs, and 3,500 poultry, altogether employing 24 permanent workers. In 1956 his capital was valued at 734,000 pesos and his annual net profit at 114,000 pesos.

Such golden opportunities no longer exist. There is not much new territory suitable for bringing into cultivation, and what is available is assigned to ejidos, in most cases organized collectively. Hence the ladder for the advancement of energetic individuals has become rickety. Faced with this situation, a considerable number of private sector mini-farmers have transformed their holdings into mini-factories either for poultry or egg production or some other specialty.

At the top end of the private farmer group mobility is almost nonexistent. Those whose farms are at or near the prescribed size limit are not permitted to enlarge their units. Those who might be purchasers think twice before acquiring an irrigation farm of 250 acres or a plantation of 500–750 acres, because they will be exposing themselves to invasion and/or expropriation. Consequently the farms in this category rarely change hands, which may be acceptable in cases where the management is efficient but is against the national interest where the management is mediocre. Altogether a situation has been reached in which the rigidities of tenure coupled with social insecurity greatly discourage the spread of good management among the private farmers.

In the ejidos the situation is worse. The only changes in occupancy which occur are those resulting from arbitrary interventions by the local bosses, a situation which provokes insecurity, not mobility. Although many ejidatarios rent their land illegally to private farmers, the authorities are trying to prevent this under threat of confiscating the ejidatario's parcela. Some ejidatarios used to expand their operations by purchasing small private farms, but this too is being officially frowned on and, if we are to believe the last census, the numbers of such cases have sharply diminished. The group is so regimented that the bad farmers cannot quit and the good ones cannot expand. So many privileges and subsidies are obtained from the government that effort has become unnecessary. No enterprise-stimulating incentives operate; nor do any disincentives. The system has become static, frozen. And this applies to more than half the nation's cropland.

The majority of well-informed Mexicans recognize that this is indeed the current situation in the farming community. But at this point they face a political and psychological difficulty. Political leaders are constantly declaring in their public statements that the Mexican Revolution is a continuing process. "We are all revolutionaries," and the warmest compliment a candidate for high office can be paid is that he is "a good revolutionary." But the Agrarian Reform was an integral part of the Mexican Revolution, one of its most outstanding features. Therefore, the Agrarian Reform is also a continuing process. To deny the one proposition would be to deny the other. This psychological dilemma largely explains the many extraordinary statements of political and agrarian leaders. For them it is necessary to find ways of keeping alive not merely the spirit but also the program of the Agrarian Reform in order to be identified as "good revolutionaries."

Not long ago a prominent minister proclaimed that there were still three million Mexicans with rights to donations of farm land. When it is recalled that according to our best estimate there are only three million farm families altogether of whom two million are ejidatarios already occupying parcelas, it can be appreciated that such statements are something of an exaggeration. Moreover, the investigation of particular groups of applicants has revealed that many of them are padded out with the names of people who are shop-keepers, truck drivers, construction workers, and even ejidatarios who already have parcelas but would like more. After all, it is only human to try to get something for nothing, and what could be more attractive than to obtain gratis from the government a productive capital asset? If it were announced that the government were going to distribute a large number of fully-equipped facilities for, say, auto repairs, there might well be three million applicants.

Very recently some signs have appeared of tentative changes in the government's policy. It has finally been admitted that the lists of land-hungry peasants are to a large extent fictional. It has also been admitted that in at least certain states of the Republic there remains no more land to donate. And it has been suggested that henceforth any new ejidatarios should be required to pay for the land they receive. These are encouraging signs; but the path ahead is strewn with political and emotional pitfalls. We should not expect from the government too much too soon.

The reader will now realize that the agricultural situation has indeed become desperate. Production has been almost stagnant for fifteen years. The old institutions of the Agrarian Reform, positive and useful as they were in their day, have become obstacles to progress. Because these have become part of the revolutionary heritage they cannot suddenly be dumped. Nor would it be wise to abandon them all; some still have an important role to play. But that does not mean a doctrine of "no change." Man survives by adaptation, and the reform institutions could and should be adapted to the needs of the present.

The remainder of this chapter will be devoted to an examination of how this might be achieved. That will involve analyzing the shortcomings of the existing legislation and making suggestions as to how the texts should be amended. It will involve a reconsideration of the concepts of ownership of agricultural land with proposals for modifying the existing tenure regimes. And it will involve trying to liberate the ejidatarios from the restrictions which beset them, without allowing rough market forces to obliterate the social values which they represent. This analysis will touch on some of the most delicate issues in Mexico's domestic politics; but once they are better understood their awkwardness rapidly diminishes, and, whatever the pessimists may say, a remodeled Agrarian Reform could be helpful to the farm community for many years to come.

DEFECTIVE LEGISLATION

It has been said, and frequently repeated by the highest authorities in Mexico, that the agrarian problem can be solved only within the framework of the law, and by applying the law rigorously and impartially. Admittedly there exists a deep-rooted tradition that local political bosses stand beyond and outside the law, that it may be dangerous to question the actions of prominent ex-politicians, that state governors and judges often act in arbitrary ways. All this, however, is only a minor part of the problem. The major difficulty lies in the ambiguity and contradictions which proliferate in the agrarian legislation itself to such an extent that even the most honest and conscientious judge would be unable to interpret the texts and apply them. Until these ambiguities and uncertainties have been removed, it would be useless to try to clear up the other component of the problem, namely the maladministration of the laws.

Let us begin with what the law has to say about the maximum limits permitted for private farms, as set out in Section XV of Article 27 of the Constitution (as amended in 1946) and in Articles 249 and 250 of the Federal Agrarian Reform Act of 1971. The language of the act exactly repeats the text of Section XV, stipulating that such properties may not exceed 100 hectares* of irrigated land, but with exceptions for certain crops, namely up to 150 ha. of irrigated cotton, and 300 ha. for irrigated bananas, sugarcane, coffee, henequen, rubber, coconut palms, grapes, olives, quinine, vanilla, cacao, and fruit trees. No definition of fruit trees is provided and it is unclear, for instance, whether the phrase does or does not include nuts and fruits which grow on bushes rather than trees. If grapes are included, then what about strawberries? If henequen, then perhaps also other agaves (for tequila), and if not why not? The ambiguous drafting of the text allows its application to be interpreted by local officials in different ways in different parts of the country, giving scope for political favoritism or victimization.

In any case the list is a most peculiar one. It is made up of a list of crops grown wholly under irrigation (grapes), others grown partially on irrigated land (sugarcane), and others grown entirely in rain-fed areas (henequen, coffee). These last should be regarded as exceptions not to the 100 ha. limit for irrigation but rather to the 200 ha. limit for rain-fed crops (see below). There seems little economic or social justification for any of these exceptions. Some are crops having an extremely high value of output per hectare, like grapes, whose value is as high as that of tomatoes (also a fruit, botanically speaking, and excluded from the list). Others, like henequen, are low value crops.

The next point to be noted is the set of equivalents to one hectare of irrigation spelled out both in the Constitution and in the Act, namely two hectares of rain-fed cropland, four of good quality pasture and eight of mountain grass or of pasture in arid zones. In any case this sentence would clearly

---

*In citations from the legislation, the figures of area are left in hectares, as in the original. One hectare equals 2.47 acres.

permit up to 200 ha. of such crops as coffee, cacao, and henequen grown without irrigation. But what is of greater interest is the provision regarding pastures. With the passing of time it was realized that the phrases used were too vague; so the act of 1971 provided that in each property the permissible amount of pasture should be determined through field studies conducted by the local representative of the Department of Agrarian Affairs. This is another instance of handing judicial powers to local officials of government, a dubious arrangement for a matter affecting so decisively the property rights of Mexican citizens.

Fortunately, it was later decided to establish a technical commission to determine pasture coefficients by districts. This commission's studies have covered the whole country in such detail that in certain states we find as many as seventy districts, each with its separate pasture equivalent, that is to say the number of hectares needed to maintain one head of cattle. The figures range from less than one hectare in areas of rich pasture up to 60 ha. and more in the arid zones. The government has announced that these coefficients will be used to determine whether or not a particular property exceeds the prescribed limit of size.

This, however, leads us to uncover a major contradiction in the texts. Having established maximum limits for ranches (400 ha. or 800 according to pasture quality), the same paragraphs of the Constitution and of the act proceed to give another definition, saying that ranches may possess the amount of grassland "necesary to maintain up to 500 head of cattle or its equivalent in other species of livestock." Combining the two definitions: 800 ha. of arid pasture with 500 head of cattle implies 1.6 ha. per animal, which is nonsense. Nowhere in the Republic do arid pastures have a coefficient as low as this. In San Luis Potosí the commission fixed coefficients of 50 and 60 ha. per head.

Suppose a man has a cattle ranch on the dry lands of San Luis Potosí with a pasture coefficient of 50 ha. per head as the average for his property. One paragraph permits him a maximum of 800 hectares which would allow him to graze 16 head of cattle, whereas another paragraph allows him 500 head of cattle which, according to the coefficient, would authorize him to possess 25,000 hectares! It is incredible that such a gigantic contradiction, which crept into the Constitution when it was amended in 1946, should continue to be repeated in the latest edition of the Agrarian Reform Act when Mexico's agronomists are perfectly well informed of the anomaly.

There are also other ambiguities. What, for example, is a "head"? This sounds a silly question, but it is of importance to a rancher whose herd fluctuates according to the calving season and the selling season. If a day-old calf counts as a "head," his herd suddenly exceeds its legal limit in the spring and falls below it in the fall. If a day-old calf is not a "head," is a month-old one, or one aged nine months? The legislation is silent, and the matter is left to the local official.

The texts talk of 500 head or *its equivalent in other species*; but nowhere are these equivalents spelled out. The United Nations Food and Agriculture Organization says that ten sheep are equivalent to one head of cattle. Does a Mexican sheep farmer then have a right to 5,000 sheep? But probably the local federal official has never heard of FAO and its equivalents, so the matter is left to his discretion. And the same principle applies to the producer of hogs or goats. He has no legal basis on which to establish his inalienability.

In a previous chapter it was explained that certificates of inalienability could be either for crop farms or for livestock farms, provided that the properties complied with the limits cited above. We further mentioned the amendment of 1971 which granted a new type of certificate to a mixed farm, i.e., one on which a rancher grows a (limited) amount of feed crops exclusively for his own use. But there is a catch in the phraseology used. It says that if a portion of a ranch is considered apt for feed crop cultivation, that part of the property becomes a crop farm for purposes of inalienability. Suppose a rancher owns a 5,000 ha. ranch on which he keeps 200 head of cattle and has a ranching certificate of inalienability. One day the federal official visits his ranch and tells him that in his opinion he has 500 ha. which could be ploughed and sown to crops. The official therefore cancels the certificate, expropriates 4,800 ha. and gives the rancher a new crop farm certificate for 200 ha. of cropland. Cases of this kind have been occurring, and the text requires amendment. Another article in the act states that an owner shall not lose his certificate through the fact of having subsequently made improvements to his property (such as introducing irrigation). If this be correct, then even less should he lose it because an official asserts that some of his land *might* be capable of improvement.

Yet another ambiguity deserves mention. The law states that a rancher may not grow crops, except in the cited case of fodder. But nothing is said as to whether a person with a crop farm certificate may keep animals. If a person has 50 ha. irrigated and then decides to build a stable and keep 100 milking cows, feeding them partly on his alfalfa and partly on purchased feed, has he acted illegally? The law is silent, but cases have occurred of officials withdrawing the certificate of inalienability for such an action.

These various ambiguities have been described at some length because they constitute one of the root causes of the juridical insecurity which prevails among Mexican farmers. Nor have we yet exhausted them. For instance, in order to obtain a certificate of inalienability an individual must, in addition to conforming to the above cited requirements, possess a legal title to his property, which rather few have because of the lack of maps and surveys. Yet there is an exception, namely when he can prove continuous occupancy in a pacific manner during the five years prior to his application. If the authorities want to expropriate his land, nothing is easier than to organize an invasion, stir up a little violence and thus demonstrate that his occupancy has not been pacific.

Probably the most vexatious question of all is what constitutes a property? If the law pretends to set size limits it must also define a "property." It is asserted that the commonest subterfuge for circumventing the law is to register properties in the names of various family members and operate them jointly as a single unit. On this subject the Constitution says nothing and the act very little. It mentions that pieces of land which are physically scattered but all belong to one person constitute a sole property. Likewise a person who owns part shares in the properties of others, has these parts accumulated in calculating the size of his property. But it is also specified that if a group of owners cultivate their land collectively or organize a marketing cooperative, their lands shall not be added together. Nonetheless, many of the large units which have been denounced and/or invaded have been family units organized in one or both of these ways.

The reader should not misinterpret. The previous paragraph was not written in defense of latifundia—far from it. The author considers that in the circumstances of Mexico at the time of the Revolution the redistribution of land was a vital necessity and that, even in the changed situation of today, the maintenance of maximum limits to the size of farm properties is entirely correct. What is wrong is to continue with a law which is so ambiguous, and in places contradictory, that it leaves the individual citizen at the mercy of unscrupulous local politicians. It is in any case desirable to transfer the application of the law from local administrative officers to the courts of law, but even honest judges could not apply the Reform Act until its text has been rewritten.

Nor do these difficulties concern solely the private sector. In the ejidos there are problems, though of a different kind. It was mentioned earlier how, when the ejidos were first formed, each ejidatario was assigned a parcela, but this parcela has no documented boundaries, and very few cultivators have obtained titles of occupancy. Theoretically he is free to leave whenever he wishes, but in practice he is a prisoner tied to his land, because, if he left, the ejido would give him no compensation for improvements he may have achieved through many years of hard work. He is forbidden by law to rent his land, even to another member of his own ejido. He is forbidden to subdivide his parcela among his sons if by doing so the new parcelas would be below the legal minimum size—which most of them would be. For any of these offenses he can be evicted.

All this, of course, dates back to the very different circumstances of 65 years ago. At that time it was highly necessary to protect the newly created groups of ejidatarios from having their land stolen back from them by unscrupulous persons. The inalienability of ejido land was a cornerstone of the new system, and successive governments ever since have been careful to maintain this feature.

But the reformers and their programs were somewhat less realistic in three respects. First, they imagined they were dealing with a world static in technology, like the medieval world whose institution they were copying, and so they made no provision for incentives to improvement and change within the ejido structure. Secondly, they overrated the capacity of human beings to accomplish decision-making in groups—local level democracy—and by endowing the ejido assembly with considerable powers inadvertently created ample opportunities for local skulduggery by the ejidos' presidents. Thirdly, they conceived of ejidatarios as identical pawns on the agricultural chessboard, believing that each and all would have the capacity and desire to cultivate indefinitely exactly the same quantity of land, and that if any deviations from conformity manifested themselves they could be suppressed.

But ejidatarios are human beings, and human beings are not like that. They have their personalities and idiosyncrasies, they have wives with differing personalities, some have large families, some have small ones. Some have a real interest in farming coupled with innate skill as a cultivator; others dislike farming, continue only because there is no alternative, and would seize the opportunity to become a truck driver. The regulations of the ejido system have tried to obliterate these differences in human nature and behavior. In all communities a considerable portion of the population accepts what is imposed from above, and among ejidatarios the more passive and inert have accommodated themselves to the system. Nevertheless, a substantial number have stronger characters, are more enterprising, innovating, active. Among these some struggle to farm areas larger than their parcelas by renting and purchasing private properties, while others who prefer and have capacity for nonfarm occupations lease their parcelas illegally and take other employment. If a system has unrealistic rules, the rules will constantly be broken.

A number of surveys have treated the topic of illegal renting of parcelas. Various figures have emerged: from 50 to 90 percent of all ejido land in several irrigation districts in Sonora and Sinaloa, from 35 to 50 percent in the Bajío, in Michoacán and Jalisco. Not only the ejidatario but also the person renting his parcela from him may be infringing the law, because this is a device by which a private farmer can operate more than the legally permitted quantity of land. Thus both of them have to conceal what they are doing.

It is often said that ejidatarios rent their parcelas only because they lack sufficient credit and technical assistance. If this explanation were valid, how is it that renting is most widespread in the better irrigation districts where credit is plentiful and the efficiency of ejidatarios relatively high? It would be nearer the mark to recognize that what is inadequate is the ejido system and its inflexibility. Many occupiers do not want to farm their parcelas—not every son wants to follow the occupation of his father; but because there is no monetary compensation for quitting, they prefer to retain their rights and make renting arrangements clandestinely. A few prefer idleness and live off their rents even if this involves a standard of living slightly lower than if they cultivated their parcelas themselves, but the majority combine renting with another occupation and so improve considerably their economic situation. Indeed, some have realized that this constitutes good business and have contrived, through applications, to obtain additional parcelas elsewhere which they also rent.

All these illegal transactions put the ejidatarios who practice them at the mercy of the presidents of their ejidos. A bribe of sufficient magnitude may persuade the local boss to turn a blind eye; but the situation is precarious, for at any moment the boss may change his mind and expose the law breakers. In many such cases the president is himself a parcela-renter, but this does not deter him: "Set a thief to catch a thief." A point has been reached where insecurity within the ejido matches the insecurity of the private farmer, an insecurity which, in different ways and for different reasons, is seriously retarding the modernization and expansion of production in that sector.

A country which maintains laws that are unenforceable invites social trouble. A head-on clash with human nature is doomed to failure. On the one hand is a series of legal texts concerning private producers so ambiguous that they are capable of innumerable interpretations, and on the other a set of restrictive regulations governing ejidatarios which do not correspond to the realities of today. As a result, to speak about "applying the law" has become in agrarian matters a sick joke. And, of course, situations such as this encourage disrespect for the law in general.

Indeed, much scope is provided for local gangsterism. Because so much decision-making which ought to be the province of the judiciary is delegated to officials of the Agrarian Reform Department and to presidents of ejidos, opportunities are created for multifarious combinations between local leaders having political posts or political ambitions. Some writers have argued that this is really beneficial in that it provides an outlet for man's antisocial instincts, which might otherwise erupt in even more unpleasant manifestations, that the political stability which has been achieved at national level depends upon tolerating, and in some instances even encouraging, political instability at local level, and that politicians can be manipulated only if they themselves are given scope to manipulate. If this were true, it would be a counsel of despair. The situation in the rural areas is certainly difficult, but it is not hopeless. Much is wrong, many people are at fault; many institutional arrangements, however valuable in the past, have become anachronistic. But surely it should be possible to find a way out of the jungle and arrive at a more lawful, socially satisfactory system of human relations?

AGRARIAN JUSTICE

Where do we go from here? In what directions do we seek solutions to these troublesome problems? A first prerequisite is to try to clear our minds of illusions, myths, and muddled thinking. A second prerequisite is to recognize that the present year is not the year 1915, and that programs which years ago were magnificent may need to be modified substantially if they are to be magnificently appropriate today.

Let us first consider the controversy about legally prescribed limits for land and animals. While some argue that these should be removed, we have to recognize that Mexico is a country with an over-populated rural community and with no present hope of offering alternative employment to all the sub-occupied people who live there. Under such circumstances it would not be socially justifiable to allow certain people to enlarge their properties inordinately while others have barely enough land for subsistence. The existing limits permit the achievement of significant economies of scale, and it is doubtful whether larger production units would register great additional economic advantage. It seems therefore reasonable to conserve this central feature of the Agrarian Reform.

When one comes to the policy of continual redistribution the situation is different. Everyone connected with farming would like to have more land, especially if he could get it for nothing, just as every citizen would like more capital. But while governments can distribute income and capital by printing money, they cannot print land. In Mexico the amount of land which in 1915 appeared plentiful is now seen to be strictly limited. Accordingly, the nation has to make up its mind whether it wants to distribute more land (a) by changing the rules of the game or (b) by applying the existing rules more conscientiously. Choosing the former would imply legislating a downward revision in the maximum permitted size of farm; choosing the latter would mean resting content with the comparatively modest amounts that would be liberated by clearing up the illegalities of the hidden latifundia.

There has been much argument as to the relative merits of these alternatives. A case can be made for saying that to allow one person (a private farmer) 250 acres of irrigation and another (an ejidatario) not more than 25 acres, offends against social justice. But to sustain this thesis it is necessary to define what is meant by "social justice." Does it imply imposing uniformity on everybody or does it signify curtailing the excesses of inequality? Most countries have opted for the latter, and in Mexico the chosen course has been that of curtailment by setting those upper limits to farm property. By the 1970s, as a result of the population explosion and the nonexistence of much new land, the argument had been partially drained of significance by the arithmetic of the situation. Unless the radical solution were adopted of allowing every farm family an equal amount of land (which in 1980 would be about 15 acres of cropland per family), a reduction of the permitted maximum from 250 acres to say 125 acres would not provide land for many additional families, and the economic losses in production would be disproportionate to the social benefits.

One conclusion which has already emerged from the present study is that it is going to be extremely difficult in the coming years to produce sufficient food for the expanding population of Mexico and that, because the amount of cultivable land is nearing its maximum, vital importance now attaches to having every acre cultivated as efficiently as possible. The national interest may have to take precedence over the private interest of the persons hoping to acquire land. Taking land from efficient producers and assigning it to persons who are not, and for many years cannot be made, efficient, may be too high a price to pay for realizing a social goal.

We have also to think of the future. A time will come when the farm population of Mexico begins to diminish and will ultimately fall to perhaps a quarter of its present size, with the consequence that the average size of farms will become larger. This transformation has been occurring, for instance, in Europe, and those countries which a century ago committed themselves to a pattern of minifundia are having great difficulty in adapting their farm size structure to the larger units appropriate to today's circumstances. It might be a pity if Mexico sought to resolve what is probably a transitory land-hunger problem by taking measures which would aggravate the difficulty of the readjustments required at a later date.

With these various factors taken into account, the balance of advantage appears to lie in resisting demogogic pressures to reduce the presently prescribed size limits, and accept the position that further distributions can be effected only by (a) eliminating illegal latifundia and mobilizing abandoned parcelas and (b) gradually opening up new areas to settlement. To announce such a policy publicly and unequivocally would require much courage on the part of the political leadership, especially after so long asserting that the Agrarian Reform is an ever-continuing process, but at this juncture in the nation's history such courage is required.

President López Portillo moved cautiously in this direction. He resisted pressures to reduce the size limits. His secretary for Agrarian Reform announced that in several states all the genuine applicants will soon have been satisfied and that in any case they will be required to pay market prices for lands granted to them, an innovation which will automatically reduce the length of the waiting lists. These are all moves in the right direction. They will do much to defuse the social unrest stemming from illusory expectations. However, they need to be followed by other measures which will tackle the insecurity of the private farmers and will introduce greater flexibility in the institutions of the ejido. In the following sections an attempt will be made to outline what these measures might comprise.

AMENDING THE TEXTS

Earlier in this chapter many examples were given of passages in the existing agrarian legislation which need to be clarified and in some instances modified. In undertaking such a cleaning-up process the first objective should be to define more clearly than at present the permitted legal limits for the size of farm properties. Let us consider first the livestock properties and the limitations regarding number of animals.

First, the word "head" has to be defined by stating the age at which a calf becomes a "head." Secondly, the texts must specify how many head in each of the other species (hogs, sheep, goats, etc.) are equivalent to one head of cattle, either utilizing the international equivalents or employing some others which Mexican experts might prefer. ("Heads" have to be defined for these classes of animals also.) Thirdly, a new paragraph is needed in the act which recognizes that herd size can fluctuate according to season. Such a paragraph could stipulate either that 500 head represents the maximum number of cattle permitted *at any season*, or that the limit may be temporarily exceeded by x percent when caused by the breeding cycle.

The other limitation imposed on livestock properties is the number of acres of grassland which they may possess. It is desirable that the reports of the Technical Commission be published and that teams of independent experts be given the task of applying the commission's coefficients of pasture to all livestock properties. This would remove the major uncertainty which hangs over the heads of cattle ranchers and other livestock farmers, enabling them to ascertain once and for all whether or not their properties conform to the limits. This matter needs to be settled quickly and finally.

Next, in respect to crop farms, and their sizes, we have argued that the central feature should be retained, namely a maximum of 100 hectares (250 acres) of irrigated land and its equivalent in other categories. However, with respect to the equivalents, while the naming of 200 ha. for rain-fed areas is satisfactory, the other two categories, of 400 and 800 ha., seem very peculiar. Both are described as "pasture," suggesting that they are not meant to be applied to crop cultivation. But if they are applied to ranches, we have shown them to be nonsense. It is reasonable that these two categories be abolished.

There remain the itemized crops for which areas larger than 100 ha. of irrigation or 200 ha. rain-fed are permitted, exceptions which grant an unjustifiable favor to a certain limited group of producers. Since it is perfectly possible to grow cotton economically on 100 ha. irrigated (rather than on 150 ha.) and coffee, etc. economically on 200 ha. rain-fed (rather than on 300 ha.), there seems a convincing case for abolishing this list of exceptions.

Certain other factors remain in determining whether a farm operator is eligible for a certificate of inalienability. For instance, in the case of persons possessing or applying for crop farm certificates, the act should clearly state what limitation is imposed on the number of animals which they may keep. Since they already enjoy an income from the crops they produce, it could be argued that the limit should be lower than that imposed on livestock farmers. On the other hand, since what is a scarce factor in Mexico is land rather than animals, there is really no case for imposing any limit at all on persons who are already restricted as to their land area. After all, in practice no restrictions have ever been applied to poultry producers who may operate as many as half a million birds.

As to the livestock certificates, the major ambiguity will be removed by determining for each farmer his pasture limits through application of the technical coefficients. There remain the concessions of temporary inalienability granted under the Cárdenas decree of 1937. The majority of these have already expired, and no new ones have been granted for many years. It would do no harm to terminate the remainder promptly.

The third category of certificates, those for mixed stock and feed farms, needs clarifying in two respects. First, the area permitted for growing fodder crops should be stated explicitly instead of by implied reference to the limits for crop farmers. While some writers advocate setting lower limits for feed areas than for general crops, it should be recognized that in many regions the carrying capacity of pastures could be significantly increased if the ranchers were permitted to grow quite a lot of feed to sustain their animals through the dry season. In view of the nation's future need for a larger volume of livestock products and the physical difficulties facing livestock expansion, it appears advisable that this latter consideration should prevail. The second change should be to delete the paragraph which empowers local officials to decide unilaterally that part of a ranch might conceivably be used for crop farming and to cancel, as a consequence, the certificate of inalienability.

Another set of desirable textual changes relates to the definition of "a single property." There are several aspects to the problem. The act counts as a single property any group of scattered farms operated under a "single administration" but fails to define this phrase. This gap needs to be filled, probably following the same rules as govern the definition of a company in commercial law.

However, one has to recognize that a farm is a much more personal enterprise than a factory, one in which several family members play a part, and some allowance should be made for this human situation. No tolerance whatever should be given to inscribing farms in the names of grandmothers, cousins, and babies, and operating them in combination as a single unit while pretending that they are separately administered by the grandmothers, etc. But perhaps the relationship of father and son should be recognized. It is suggested that the texts should permit a farmer to purchase or rent a piece of land (up to the maximum size in each case) for each of his adult sons—or, if preferred, up to a maximum of three sons—and treat these properties as separate administrations even though in practice they be worked jointly, provided that all work full-time in farming. This would go far to satisfy a very natural human desire and would not give rise to any great number of latifundia.

All other groupings of family members' properties should be strictly additive for calculation of size of enterprise, unless separate administration could be convincingly proved. Furthermore, it should be stated clearly that any land which is rented, if only for one season, should count as part of the property. Thus a man who owns 150 acres of irrigated land would be permitted to hire an additional 100, whereas a man who already operates 250 acres would not be able to hire further areas.

It will be noted that in the event of the adoption of some or all of the changes advocated in preceding paragraphs, corresponding amendments would have to be made in Article 27 of the Constitution. The whole topic is therefore one which deserves the most careful consideration before action be taken. The central purpose should be to establish in precise legal language what is allowed and what is not allowed in regard to agricultural properties and with respect to certificates of inalienability, thus removing not only the ambiguities but also the temptation for local officials to interpret the texts in favor of their friends and to destroy people whom they happen to dislike. Indeed, an essential corollary of these legislative revisions would be to remove jurisdiction in agrarian matters from the Secretary of Agrarian Reform and his officials and place it firmly in the hands of the judiciary. But the courts of law can be requested to administer the interpretation of this legislation only when it has been redrafted in a manner which enables judges to apply it.

LAND TITLES

To clarify the legal texts solves only part of the agrarian problem; another part is to provide documentary basis for security of tenure. It will be impossible to establish peace in the farm community until all those who cultivate land or raise animals have obtained pieces of paper showing the nature and extent of their entitlements. It has been estimated that at the present time less than 10 percent of all ejidatarios have titles of parcela occupancy, and no one knows how many private farmers have legalized titles—certainly more but not much more.

This, of course, is a quite common situation in a developing country; complete cadastral surveys accompany a later stage of sophistication as does a larger supply of surveyors. But what differentiates Mexico from other developing countries is her government's attempt to legislate in detail the size of agricultural properties, an attempt which becomes abortive unless it can be implemented with the help of documentary land titles. For this country the provision of titles is not a luxury but a necessity. A landowner without a title cannot be granted a certificate of inalienability, and let it be hoped that soon everyone will have his certificate of inalienability. The two documents should accompany one another, and until they do so the evasions, the invasions, and other illegalities will not be eliminated.

At various points in this study it has been urged that discrimination against ejidatarios should be terminated, and this is particularly true in respect to their documents. Why should inalienability be a privilege reserved for private farmers? Why should not ejidatarios be granted titles (of occupancy) for their parcelas? Why should they be given merely certificates which entitle them to land but not to any particular piece of land? A number of arguments have been advanced in justification of the present state of affairs.

One is that to grant to each ejidatario a title to his parcela would be to move in the wrong direction, i.e., away from the socialism which should be the goal of ejidos. But on this point the socialists should explain themselves. There are some who take the view that collectivist organization is so "correct" that it should be imposed irrespective of peoples' wishes. Others, however, believe that collectivization should be voluntary and admit, furthermore, that collectivization by persuasion will be a slow process. In such case, there will be for many years to come two kinds of ejidatarios, the collectivized and the uncollectivized. Why should the latter be discriminated against? Why should they be treated worse than private farmers in the matter of titles? After all we do not deny deeds of ownership to city dwellings on the excuse that one day the nation's stock of dwellings may be nationalized.

Another argument often advanced is the alleged physical impossibility of the task: There are two and a half million farm families covering 350 million acres of land, and it would not be possible to provide enough surveyors to establish the boundaries of all the farm units in this vast area. This argument is founded on an exaggeration and a misrepresentation. Of course, if one were faced with 350 million acres of virgin territory hitherto uninhabited, and were given the task of drawing the boundaries for three or four million separate properties, the difficulty in the short run would be insurmountable. But Mexico is not like that; it is not virgin territory. On the contrary it has been settled for centuries, and although as a result of the Agrarian Reform the pattern of occupancy has greatly changed, the vast majority of operators know what land is theirs, even without documents, and recognize what are their neighbors' boundaries.

Much publicity has been given to land wars within or between certain villages, in which sometimes people get killed, but these cases constitute a tiny fraction of the whole. It would be a reasonable guess that 95 percent of the farm units in the Republic have boundaries on which no disputes arise. Accordingly, to all of these it would be a simple matter to grant titles while the surveyors would deal with the disputed cases. Nor should it be impossible to augment fairly rapidly the number of persons needed for this task; the qualifications for a *auxiliary* surveyor do not have to be particularly high. This would be an activity in which the army could perform a helpful role.

What about the content of these land titles? It is suggested that they should be extremely simple, all of them following a uniform pattern and limited to a minimum number of essential points. In considering what the contents should be, it may be useful to remind ourselves of the recent evolution of attitudes with respect to property in land. Whereas during the Middle Ages and in earlier times property was conceived of as belonging to the king or tribal chief or community, who assigned to individuals certain specified rights of property usage, in the eighteenth and nineteenth centuries the concept changed to one of unfettered and absolute individual rights.

However, this extreme individualism did not last long, and has been overtaken by a new attitude which accepts a wide variety of restrictions on the use of land imposed by government for the public good—restrictions concerning the type of buildings permitted in certain zones, the discharge of noxious effluents into rivers, and other matters relating to the environment. In short it is currently accepted that rights in private property are not absolute but can and should be limited by legislation.

It is suggested that in Mexico this modern concept of property be reflected in the language employed in the title to land. This title should be drafted in the form of a contract between the State on the one hand and the individual on the other, in which the latter is accorded certain specified rights. In this context one may refer to the opening sentence of Article 27 of the Mexican Constitution which says that the ownership of all lands and waters within the national boundaries was originally vested in the nation, which has the right to transfer the dominion of these to individuals thus constituting private property. The sort of language we are envisaging would constitute only a modest extension of the idea enshrined in this sentence. The State would be designated once again as the ultimate owner and would transmit to individuals certain specific privileges. The title would be a negotiable legal instrument, in the sense that its owner would have the right to sell it or lease it to another person in the same way that an owner sells or leases property today; but in this case of farm properties provided always that an individual could not, by owning or leasing, become the operator of more acres than the legislation allows.

The title owner would have the right to mortgage his title to obtain credit, and in the event of default the title would pass to the bank, which would be empowered to dispose of it to another person whose land holding was not yet at the legal limit. These titles or contracts between the State and farm operators would have a minimum duration of, say, thirty years, renewable at the option of the operator. This should provide a period long enough to encourage investment in land improvement works, an incentive which today is conspicuously lacking. Furthermore, since the document would have the form of a contract with the State, its possession would obviously be equivalent to recognition by the State of the inalienability of the said property. Indeed such a title would be granted only to persons fulfilling the conditions of inalienability, and therefore would also serve this purpose, rendering unnecessary the issuance of a separate certificate. Security of occupancy would be guaranteed by the State's signature appended to each operator's title.

FLEXIBILITY FOR EJIDATARIOS

Because the ejidatarios are confined in a wholly unrealistic web of restrictions they are persistent lawbreakers. There are far more ejidatarios breaking the law in regard to their parcelas than there are neolatifundios. Social justice requires that sooner or later, and preferably sooner, these people should receive treatment as good as other members of the farm community.

It has been urged that the private farmers have land titles which in form and content give them a guarantee of inalienability. Exactly the same reasoning applies to ejidatarios, and it is equally desirable that they should have titles too. Ideally and ultimately the form of title should be the same for both groups of cultivators. In a philosophical sense all land is held in usufruct, and the thesis proposed here is that whatever rights the State decides to concede for the farming use of land should be granted equally to all.

However, we are not starting from a tabula rasa but from an existing situation in which it would be unwise and unrealistic to leap immediately into a uniform system of tenure. The ejidatarios have too long been cocooned and protected from the pressures of the market, too long imprisoned so that if, like the prisoners in *Fidelio*, they were suddenly brought out into the sunlight, they could not see where to go. Some transitional arrangements will have to be made. The land titles of ejidatarios should perhaps contain certain additional clauses valid for a limited number of years.

The title should undoubtedly recognize the right of renting or, in more exact language, the right of the parcela occupier to assign his contract (with the State) to another person. But in the case of the ejidatario it might be advisable to stipulate that he may not rent his land for a period longer than, say, two or perhaps three years, giving him the opportunity at the end of the lease to review his position and decide whether to continue renting or resume occupancy himself. This would protect those in a weak bargaining position from being pressured into alienating their land against their better judgement. A corollary of this limitation would be a clause forbidding the outright sale of an ejidatario's title. It should be underlined that the maintenance of these restrictions can be justified for a transitional period only; as soon as feasible they should be removed and the principle of social and juridical equality put into practice.

Some authors with reforming intentions have suggested that the ejidatario should indeed be allowed to rent his parcela but only to another member of his own ejido. This seems an undesirable concession to the old system, and is tantamount to saying that prisoners may talk to one another but not to visitors from outside. In reality the ejidatario would be provided a considerable degree of protection against unwise action, because it needs to be remembered that the arrangement proposed here would require all renting of titles to be registered with a notary public or other legal authority, if only in order to ensure that the lessee was not thereby increasing his land holding beyond the legal limit. And this last would be the only valid ground for refusing to sanction a leasing arrangement.

It should be added that these titles, just like those of private farmers, would be documents usable as collateral for obtaining agricultural credit—an aspect examined in more detail in our next chapter. This would relieve the ejidatario from the onerous restriction of having to form a cooperative credit society before becoming eligible for credit.

To a considerable extent the provision of land titles would also overcome another defect of the present system, namely the disincentive to improve the productivity of the parcela. If an ejidatario planted, for instance, fruit trees he obviously could let his parcela at a higher rent than if it were merely bare land. Similarly when after the transition period he obtained freedom to sell his title, any improvements he had made would be reflected in the price he obtained from the purchaser, thus giving him a real incentive to invest.

Nonetheless, while the lessor would be taken care of, the same would not be true for the lessee. Suppose a man rents a piece of land, in this instance from a private farmer, for a long period of ten or fifteen years. He is not going to invest in any capital improvements during that period unless he is assured of appropriate compensation at the end. Consequently it is necessary that all leases of farm land contain a clause giving the lessee the right to claim in the courts at the termination of his lease full compensation for improvements he has effected, a right supported by the provision that he cannot be evicted until the compensation has been agreed and paid by the lessor. Such a clause is a commonplace in farm leases in those European countries where such leasing is widespread.

It is reasonable to envisage that the above proposals would go far to creating flexibility in the ejido system and to removing, after a transition period, the vexatious discriminations to which ejidatarios are subjected. Such a bold adaptation of the institutions of the Agrarian Reform is, however, likely to encounter opposition in two quarters. First, there are those who, far from wanting to liberate the ejidatario, wish to tighten the organization of the ejido through collectivization. Our discussion earlier in this chapter of the pros and cons of collective farming, although emphasizing the disadvantages, has not excluded the possibility that some ejidatarios may establish collectives really voluntarily and that these units or some of them may survive. Thus it seems likely that in the coming years two forms of ejido organization may exist side by side, which would be no bad thing since this would help to demonstrate their respective merits. But for the uncollectivized ejidos, the great majority, their progressive liberalization is the key to improving their technical efficiency and overall productivity. The provision of titles to these ejidatarios (and to comuneros) with the rights and privileges and temporary restrictions outlined above should become a basic component of the remodeled agrarian reform.

There are others who fear that to introduce so much flexibility would destroy the ejido as an entity, and after all the ejido was the main pillar of the original Reform. It is important here to clarify the meaning of words. A careful reading of the early writings shows us that the reformers expected the ejido to act as a unit of social, cultural, and economic organization, providing for its members services in all these fields. Yet it is widely agreed that, with rare exceptions, the ejido has performed none of these functions. At best it has operated as a vehicle for political control by the government party, for getting out the votes at election time. At worst it has facilitated the domination of the ejidatarios by the local mafia. What the proposed flexibility might break is not the ejido, which has never existed except on paper, but the power of the bosses.

For there remain many tasks which the ejido could and should be performing. One of these is to organize the cooperative purchase of material inputs, of seeds, fertilizers, fuel oil, fencing materials and so on, required by its individual members. Another would be to organize the grading, packaging, and marketing of members' produce. These are real jobs for a grass-roots organization grouping a few dozen or a few hundred mini-farmers. If local leadership and energy could be oriented to the provision of services such as these, the ejidos would run no danger of being weakened. On the contrary, they would become much stronger organizational units than ever they were in the past.

We can also expect to hear objections to our proposals for the private farm sector. There are some who might describe them as a deceitful device for nationalizing farm land and who would allege that the private farmer would be surrendering his land to the State without a cent's worth of compensation. This is a specious contention, which fails to take account of the actual situation in Mexico. The private farmer may think he owns his property, but he stands exposed to invasion and/or expropriation at any moment. In exchange for this nagging insecurity he would be offered a contract valid for thirty years which guarantees to him, over the signature of the government, inalienable possession of his lands and animals, a contract which he can assign or sell just like any other title. Under existing circumstances this represents a more advantageous deal than any private farmer could in his rosiest dreams envisage.

Reviewing the tenure situation in its entirety, we reach the inescapable conclusion that the old Agrarian Reform has come to the end of its usefulness and has instead become a bar to progress in farming. That does not mean that it should be thrown away completely. It contains valuable elements which should be preserved; others which need to be modified. Given sufficient courage to adapt its institutions to the realities of today it is still capable of long and useful life. The proposals put forward here attempt to deal with each of the main problems in a pragmatic manner. They call for a tapering off and final termination of land redistribution; for revisions of the legal texts to make them intelligible and applicable; for ruthless action against hidden latifundia (however politically influential their owners); for a universal and uniform system of land titles for private farmers and ejidatarios (which would solve the problem of inalienability and provide a realistic basis for promoting the much-needed long-term investment); for gradual relaxation of the restrictions affecting the freedom of action of ejidatarios so that in the not too distant future their inferior juridical and social status would be obliterated; and finally for the transfer of the power of judicial decision-making in tenure matters from the executive branch where it is misplaced to the judiciary where, both at the national and local levels, it properly belongs.

It seems reasonable to hope that such a package deal would provide a solid base for peace in the countryside. It would rapidly eliminate the too widespread law breaking by both private farmers and ejidatarios, as well as the capricious and arbitrary interventions of local officials of the executive. By substituting a legal framework for the present anarchy the justification, if ever there was one, for invasions and other acts of violence would be removed. Under the new circumstances, the government would have every reason for punishing such acts severely.

President López Portillo is reported to have said in a speech shortly after his nomination as candidate for the presidency: "We are making an appeal for greater production, and it would ill behove us to refuse to provide the juridical basis which encourages and makes that possible." A mature society has to operate under a system of law not of arbitrary interventions. The law has to be applied objectively, and its application has to be seen to be objective. The farming community deserves to be granted a larger measure of social justice, but social justice has to be based in social realism. It is perfectly possible in Mexico to move toward a homogeneous rural social structure in which all receive equal treatment and enjoy the same freedoms under the law. All that is necessary is the political will to accept and implement the measures which will achieve these goals.

9. Farm Finance

Modernized tenure legislation would provide much of the basis for entre-preneurial motivation which has been conspicuously lacking in most of the Mexican farm sector. Furthermore, it would ensure through the market mechanism of land transfer that the farms gradually passed into the hands of the more competent operators. While this might be called the institutional environment in which farming functions, another important element is the economic environment—in particular, agricultural investment and the facilities for increasing it and the evolution of the prices of farm products as an inducement to expanding output. In what are now the industrialized coun-tries, before the State began assuming responsibilities of economic manage-ment, the necessary inducements to opening up new land, to intensifying the cultivation of old land, and to meeting the requirements of consumers, were provided through the free play of interest rates and commodity prices. But even in industrialized countries that has ceased to be the case in recent years. And perhaps it never was the scenario in such developing coun-tries as Mexico.

Due to Mexico's agrarian reform, more than half the crop area (that cultivated by ejidatarios and comuneros) was in effect removed from the market economy. Neither this land, nor the pastures and forest belonging to ejidos and comunidades, could be bought or sold. Moreover, since this land was not held in outright ownership, and could not be offered as security for loans, special arrangements had to be made to provide its operators with farm credit. In addition, partly as a result of unhappy experiences in world markets during the depression of the thirties, postwar governments in most countries have pursued policies of price support and/or guarantee for a wide range of farm products, a system which Mexico also adopted from the sixties onwards.

We thus have in Mexico a largely man-made economic environment for the farm sector, and the theme of the present chapter is to observe how this has evolved, what are its advantages and disadvantages, and what might be done to make it more propitious. The analysis falls into three parts: first, an examination of farmers' investment in their land, their animals, and their equipment; secondly, an analysis of how farm credit has functioned as an instrument partly to stimulate investment and partly to facilitate orderly mar-keting; and, thirdly, agricultural prices as a reflection of market equilibrium and/or of government intervention policies.

FARM CAPITAL

Agricultural capital is not merely money; it is the land and animals, the machinery and buildings, etc. used in production. Public agricultural investment is the undertaking of irrigation works, the establishment of a research station and so on. A farm sector which has a functioning system of investment and is adequately capitalized enjoys the prerequisites for expansion and for the bettering of farmers' incomes. Where the average unit of farm manpower is associated with a large quantity of productive agricultural capital, average farm income will be high. Indeed, it is the abundance of such capital which chiefly distinguishes a modern and prosperous agriculture from a primitive and traditional one; thus our task is to examine how the Mexican situation has developed in respect to these factors.

The items which are conventionally classed as agricultural capital include land, livestock, buildings, irrigation works, machinery, vehicles; and also stocks of production materials and of unsold output, cash reserves, and a number of miscellaneous components. Apart from the statistically advanced countries, very few possess information on these latter items of working capital. Even with respect to fixed capital, information is normally collected only periodically through farm censuses and its reliability varies greatly from country to country.

In Mexico the most recent information derives from the fifth farm census undertaken in 1970 and refers to the year 1969. In absolute terms the data have changed greatly since then, as a result of the devaluation, the continuing inflation, and other factors, but in relative terms—the different classes of capital and the level of capitalization of different types of farms—the situation has probably changed very little.

The principal components of farm capital recorded for the year 1969 were the following:

	million pesos
Land	89,326
Livestock	64,798
Buildings	4,819
Irrigation works	2,327
Machinery and Vehicles	8,511
Total	169,781

Land thus accounted for more than half of all agricultural capital, and livestock for more than one-third. The remaining items together amounted to less than 10 percent of the total. As will become apparent, there are reasons for suspecting that the farm land was somewhat undervalued by the informants and the livestock overvalued, but these defects would not appreciably change the broad picture.

Land Values

In normal circumstances the price of farm land, as of any other capital asset, should reflect its income-producing potential. However, in almost all countries the market value of agricultural land exceeds its economic value, often by a considerable amount. One reason, especially prevalent in unindustrialized countries, is the land hunger of the rural population. For the great majority of rural people the cultivation of land is their only source of income; they have no alternative, and so compete ferociously for the insufficient amount of land available.

Another reason is the modern phenomenon of inflation, which complicates the problem of how to calculate the long-term yield of a productive asset. In practice, land tends to be regarded as a sound hedge against inflation, in the sense that, because governments generally have a commitment to ensure to their farm people a minimum standard of living, it can be safely assumed that agricultural prices will be adjusted to take account of the depreciating value of the currency, and that consequently the productivity of farm land will not decline in real terms; indeed it may be expected to increase as a result of the application of improved technology.

A further influence which keeps farm land values at relatively high levels, especially in developed countries, is the "prestige factor," the social status which attaches to the possession of a tract of land. Industrialists, stockbrokers and the like acquire agricultural properties, even though they may operate them at a loss (which perhaps can be offset for tax purposes). Many a small farmer prefers to own a piece of land, even though were he a tenant he would have more capital available for expanding his business. Farm families that decide to migrate to the cities seldom sell their farms, preferring to lease them to other operators, thus retaining something to fall back upon in some future time of adversity.

The influence of all these components of land value varies greatly according to the quality of the land and its location. A property from which two or even three crops can be harvested each year is obviously worth more than land which can support only one head of cattle per 100 acres. Land which is within commuting distance of a city and is situated in an attractive landscape has higher amenity and prestige value than land in distant, semidesertic locations. For lack of space, the statistics we shall be citing refer to average farm land values over a whole Mexican state, which inevitably conceals wide variations between the micro-regions within each state.

Moreover, Mexico has another problem peculiar to itself, namely the ejido lands which theoretically possess no market value since they cannot be bought or sold. The census authorities overcame this difficulty by deciding to impute to ejido land a capital value equivalent to the value of similar quality land in private ownership in the same district.

From Table 9.1 it may be noted that the national average value of crop land was 83.35 dollars per acre in 1969, for all states and all types of farm. The range between states extended from a low of $33 per acre in Yucatán to a high of over $280 in Morelos. Generally speaking, the most influential factors appear to be soil fertility and climate, good or deficient rainfall and the presence or absence of irrigation. If we quoted figures for individual *municipios*, these factors would stand out still more clearly.

When analyzed according to type of tenure, the highest land values are

TABLE 9.1

**AVERAGE VALUES OF ARABLE LAND BY REGIONS
AND BY TYPE OF FARM: 1940, 1960, AND 1970**

(in dollars per acre)

	1940	1960	1970
Region			
Northwest	30.20	89.76	127.21
North	26.64	48.62	72.41
Northeast	25.25	62.93	73.48
North-center	11.82	32.92	55.71
West-center	23.60	47.39	79.47
Center	37.32	81.51	136.50
Gulf-south	24.70	45.64	68.85
Peninsula	20.59	75.78	34.80
Type of Farm			
Ejidal, average	23.89	59.11	84.19
Private, over			
12.5 acres	25.25	54.51	79.40
Private, under			
12.5 acres	34.05	66.88	121.84
National average	25.54	57.13	83.35

SOURCE: Statistical Office, decennial farm censuses.

NOTE: Values in real 1970 terms. The 1940 and 1960 figures were multiplied by the overall Mexican inflation index up to 1970. In 1970 the exchange rate was 12.50 pesos to one dollar.

recorded for the private farms of under 12.5 acres and the lowest for those over 12.5 acres. The average value of the ejido and comunidad land was much closer to that of the larger than that of the smaller private farms, which is surprising since the average size of the ejidatario plots is just 12.5 acres. The fact that ejido plots were priced on average at 30 percent less than private small plots implies either that ejido crop land was of poorer quality or that the census authorities did not realistically apply to it its market value. Perhaps both factors played a part. In any event the value differences according to farm tenure and size were much smaller than those between states, which suggests that the climatic factors were preponderant.

The extent to which land values in Mexico reflect land productivity can be verified by comparing the two elements (Table 9.2). The figures show that at the time of the census the average value of crop land was roughly equal to three years (gross) output on mini-farms and ejidos, but to less than 2½ years gross output on the larger private farms. However, the average for the center region was 4½ years output, and even 5½ years for ejidos (rising to 8½ years in the state of México). By contrast the ratio is low in the northern states and lowest of all in the northwest.

Do these contrasts reflect a greater land hunger in the center, or some other influences? Undoubtedly, demographic pressure is severe in the rural areas of this region; farms are particularly small in the state of México. Yet another factor is also at work. Land values normally reflect the value of *net*

TABLE 9.2

ARABLE LAND VALUES
PER THOUSAND DOLLARS OF CROP OUTPUT PER ACRE: 1970

(in dollars per acre)

Region	Private Farms over 12.5 Acres	Private Farms under 12.5 Acres	Ejidos & Communities	Average
Northwest	1,637	1,712	2,354	1,956
North	2,538	2,507	2,948	2,736
Northeast	2,052	2,287	2,561	2,224
North-center	2,961	2,368	4,658	3,740
West-center	1,747	2,267	2,821	2,344
Center	3,246	4,160	5,533	4,503
Gulf-south	4,321	1,920	2,392	2,989
Peninsula	2,793	1,538	2,223	2,370
Mexico	2,405	3,006	2,929	2,698

SOURCE: Statistical Office, farm census of 1970.

output for which concept no reliable figures are available, and the difference between net and gross output is by no means uniform over all the Republic. In the northwest, where purchased inputs amount to 30-40 percent of the value of gross output, the net output figure is correspondingly much lower than the gross, whereas in the more backward center region the difference may be only 5-10 percent.

And yet by international standards farm land in Mexico remains incredibly cheap. Some figures published by FAO show that around 1970 the only important country with farm land values resembling those of Mexico was Canada; elsewhere values ranged from $160 in the United States to $2,400 per acre in Belgium and Japan, or thirty times the Mexican figure.* Of course the situations in these two latter countries are totally different: mini-farms predominate even more than in Mexico, net output per acre is much higher, interest rates are lower inducing investors to accept lower rates of return on agricultural investments, and the prestige value of owning farm property is more in evidence.

To have low-priced farm land is not necessarily an advantage. At the level of the individual farmer it may discourage investment in improvements because of the limited possibility of recovering such outlays in terms of the land's selling price. At the national level, the fact that land is cheap creates the illusion that it is plentiful, whereas good, cultivable land is extremely scarce. Such a misconception tends to have harmful influences on farm policies.

*Food and Agriculture Organization of the United Nations, *Agricultural Adjustment in Developed Countries,* Rome, 1972, pp. 77. Since that time there occurred a boom in farm land values in the U.S.A, while inflation raised those of Mexico to a presumed $150 per acre at the end of the seventies; but broadly the contrast remains.

It is difficult to ascertain how long this state of affairs has persisted in Mexico, because the land value statistics of earlier censuses are fraught with ambiguities and errors. We may conclude tentatively that values have increased just slightly less rapidly than the rate of growth of gross crop output per acre. If we bear in mind the process of modernization, which although gradual, requires an increase in inputs, the net output will have grown more slowly than the gross; hence an apparent concordance has been maintained between net output and land values. One startling occurrence has been the rapid rise since 1960 in the value of the private mini-farms, which may be explained by the fact that their number was declining sharply while the demand for them as part-time units remained strong.

Livestock Capital

Whereas the value of farm land is determined partly by its productivity and partly by other factors, that of farm animals is more directly related to what they produce. The prestige factor enters only with respect to horses and other work animals. That is, a farmer with four horses is more respected than one with two, though he needs only two for the farm work; equally at a lower economic level, a man with four oxen rates higher than one with merely two. In a country such as Mexico, where farm operations are only partially mechanized, this factor still plays a role.

At the last farm census the capital value of all farm animals was recorded at just over 5,000 million dollars (at the then rate of exchange), and of this amount cattle accounted for 70 percent. Of these cattle about one tenth were classified as *fino*, meaning belonging to some recognized breed, and were worth almost twice as much per head as the others. In 1970 the average value per head of all cattle was 175 dollars, of hogs 31, of sheep 14 and of horses 97. To bring these figures up to 1980 values we should multiply them by a factor of 2.7 to allow for Mexican inflation and the devaluation of the peso.

The distribution of livestock capital between the different types of farmers was as follows: private farms over 12.5 acres, 43 percent; private farms below 12.5 acres, 17.5 percent; ejidos and comunidades, 39.5 percent.* The ejidatarios and comuneros had a larger share in *numbers* of livestock, but their animals were of poorer quality on average and were valued at 20-25 percent less per head. They also suffered another disadvantage, namely that, because their farms were less mechanized, a much higher proportion of their stock was in work animals rather than productive ones.

The value per head of most classes of livestock was lowest in the northern regions and highest in the center, which at first glance appears strange, considering that the former are the progressive farming areas. In the case of cattle the explanation is that in the north the herds have a lower average age, since the animals are destined for sale to feedlots (at home or in

---

*For reasons explained in the Appendix, of the livestock registered as "in centers of population," 25 percent were alloted to small farms and 75 percent to ejidos.

the U.S.A.), whereas dairy herds predominate in the center. The hog and poultry production of the center region is more commercial and intensive than that of the north, where it is mainly a subsistence activity. Horses and mules have a higher value in the center where they are used mainly as work animals.

If we examine the historical evolution of livestock capital, we would expect it to reflect the increase in animal numbers and in their productivity. However, according to the censuses this capital has increased significantly faster than these two factors in combination. For instance, between 1950 and 1970 the number of animal "units" in the whole country increased by 48 percent, while the productivity per "unit" increased by about 31 percent. Together these changes would justify an increase of some 94 percent. Yet the recorded increase (at 1970 prices) was more than fourfold, the difference being particularly high in the case of cattle. It is true that the price of beef increased during that period some 40 percent more than prices in general, but that would not be sufficient to account for the recorded rise in animal values. Moreover, the prices of other meat and of milk remained rather stable in real terms, while that of eggs fell.

We are forced to the conclusion that either livestock capital was under-reported in 1950 or that it was overvalued in 1970 or both. This anomaly was apparent equally in the ejidos and the larger private farms, but less so in respect to the animals kept in centers of population. It was least noticeable in the value of work animals, which is only natural since their productivity is hardly capable of much increase.

It is also instructive to relate livestock capital to the area of farm land operated by the different types of farms. Whether this be calculated on the basis of crop area or of total agricultural area (including pasture) some distortion is inevitable, since private farmers and ejidatarios do not each have the same ratio of crop to grass land. However, in Mexican circumstances, probably the former is the more meaningful. Thus while in 1950 there was not more than 20 percent difference in the livestock capitalization per acre of cropland between the larger private farms and the ejidos, by 1970 the difference had widened to more than threefold. This resulted mainly from the private farmers effecting a substantial increase in the number of animals kept per 100 acres—their reaction to having to surrender large areas for the establishment of new ejidos. The ejidos, although acquiring several million additional acres of pasture, did not succeed in increasing their stocking densities.

A more surprising development was the intensification of livestock production on the small farms of under 12.5 acres. Between 1950 and 1970 the value of livestock capital in this group increased more than tenfold (at 1970 prices) while the gross value of their livestock output rose nearly threefold. (It seems likely that the wide discrepancy between these two multipliers was caused by a serious underreporting of production in 1970.) There is no inherent reason why ejidatarios with 10-12 acres should not have developed livestock activities—dairying or hog or poultry production—on a scale similar to that of the small-scale private farms. Credit was available from the State banks, as we shall see, yet the necessary enterprise was lacking, as well as adequate security of occupancy of their lands.

The overall impression gathered from this analysis of livestock capital is one of sluggish growth, with exception of the private mini-farms, and of insufficient investment in improvement of both the number and quantity of productive animals. The larger scale farmers, or ranchers, could have afforded to augment their herds and improve the quality of their stock, but in so doing they would have run foul of the agrarian legislation and become liable to expropriation. Instead, they preferred to retain their properties and refrain from improving their animals. The ejidatarios lacked access to sufficient feed supplies, because on the one hand the ejidos did nothing to better the carrying capacity of the grazing areas which were and are held collectively, and on the other it was uneconomic to expand production on a basis of purchased feeds at the prevailing level of feed and meat prices unless they could enter the big league of commercial poultry production, which they had neither the resources nor the know-how to contemplate.

Other Agricultural Capital

The remaining components of agricultural capital are chiefly machinery, implements, vehicles, pumps, irrigation equipment, buildings, and other constructions. The most mechanized regions of the country are the northwest and north, where at the 1970 census machinery and vehicle capital attained 30 to 40 dollars per acre. The least mechanized were the states of the Yucatán peninsula plus Oaxaca and Guerrero. Irrigation works nowhere register a high figure since most of these are federal property.

During the twenty years between the third and the fifth censuses the capital value of these various items in combination roughly doubled in real terms, but the increase was concentrated on the private farms. Thus the value of machinery and vehicles per 100 acres of crop land roughly trebled, in real terms, on the larger private farms, sextupled on the smaller ones, but increased by only 9 percent on ejidos.

In Chapter 6 we found that in 1970 the mechanized proportion of the crop area was 25.7 percent for larger private farms, 4.3 percent for smaller ones and 13.0 percent for ejidos. Yet at the same date the per acre value of machinery capital for each of these three groups was 42, 220, and 18 dollars respectively. It would appear that the private farms of under 12.5 acres are overmechanized, as they are in many other countries, especially as they also spend considerable sums on machinery hire. The ejidos and comunidades appear undersupplied with machinery, even in relation to the relatively small proportion of their crop land that is mechanically cultivated. But ejidatarios also tend to hire machinery on occasion, and furthermore a considerable part of their land is rented to private farmers and cultivated by machines belonging to the latter.

It can be argued that in view of the overpopulation prevailing in the majority of ejidos, rapid further mechanization would be an uneconomic policy, merely displacing manpower already underemployed. This argument is partially valid but, as pointed out in Chapter 6, the retention of a large number of work animals means devoting many scarce acres to feeding them, acres which could be growing cash crops. Certainly the decision-making of the operators of private mini-farms has been in favor of mechanization.

Total Farm Capital

Farm capital, considered as a whole, has been increasing in real terms from 1940 onwards. Between 1940 and 1950 it roughly doubled in real terms, and doubled again between 1950 and 1970 (see Table 9.3). The annual growth rate has been progressively declining, and this slowing down apparently persisted during the seventies. Apart from the steady rise in the importance of machinery, the shares of the other components remained rather constant from census to census until in 1970 we encounter the above-mentioned anomaly of the land values being apparently too low and the livestock ones too high. If these figures are true, it would imply a decapitalization of farm land prior to the 1970 census, a proposition difficult to entertain.

Relating total farm capital to the agricultural area (crop plus pasture), we find its value increased some 68 percent in real terms in the twenty years to 1970, and it has certainly continued to increase since then. Capitalization is particularly high in the states adjacent to the huge market of Mexico City (such as Morelos and the state of México), due mainly to the high values of farm land in those locations.

If we calculate farm capital per operator, there emerges a large contrast between private farmers and ejidatarios. In 1970 the figure for private farmers

TABLE 9.3
CLASSIFICATION OF FARM CAPITAL: 1930 TO 1970

	1930	1940	1950	1960	1970
	Millions of Pesos				
Land	2,288	2,781	20,684	63,444	89,325
Livestock	754	1,140	6,205	21,054	64,798
Buildings	183	222	584	1,762	4,819
Irrigation Works	126	101	522	1,922	2,327
Machines & Vehicles	71	142	1,244	4,605	8,511
Total:	3,422	4,386	29,239	92,787	169,780
	Percentage Distribution				
Land	66.9	63.4	70.7	68.4	52.6
Livestock	22.0	26.0	21.2	22.7	38.2
Buildings	5.3	5.1	2.0	1.9	2.8
Irrigation Works	3.7	2.3	1.8	2.1	1.4
Machines & Vehicles	2.1	3.2	4.3	5.0	5.0
Total:	100.0	100.0	100.0	100.0	100.0
	Millions of 1970 Pesos				
Land	34,663	27,810	62,052	90,598	89,325
Livestock	11,423	11,400	18,615	30,065	64,798
Buildings	2,772	2,220	1,752	2,516	4,819
Irrigation Works	1,909	1,010	1,566	2,745	2,327
Machines & Vehicles	1,076	1,420	3,732	6,576	8,511
Total:	51,843	43,860	87,717	132,500	169,780

SOURCE: Statistical Office, decennial farm censuses.

(large and small together) was 12,500 dollars per operator, for ejidatarios it was 3,000. These differences were of the same order of magnitude in all regions of the country. And they were equally notable in all the components: in land value, in livestock capital, in machinery, buildings and other equipment. Of course, this was largely explained by the larger size of many of the private farms. However, if we consider only the private farms of under 12.5 acres, which in size are entirely comparable to those of the ejidatarios, the former have nearly 50 percent more capital per operator. Moreover, almost two-thirds of the ejidatarios' capital consisted in the (imputed) value of their land, but only 42 percent of that of the private mini-farmers, leaving them a higher proportion for operational purposes.

Indeed, apart from their land and an insufficient quantity of livestock, the ejidatarios possess little capital. In the central and southern regions other capital constitutes less than one percent of their investment. All they have are rudimentary hand tools and machetes, no buildings for storage of produce or what little fertilizer they may buy, and no motorized transportation since they rely on their donkeys for taking produce to market.

Before leaving the topic of agricultural capital, it is of interest and importance to attempt to measure its productivity and the evolution of that productivity through time. This is a hazardous procedure because of the defects in the capital statistics and still more in those of livestock production previously referred to. However, with all due reservations as to the reliability of the data, and using our own preferred higher figures of livestock production, it emerges that between 1950 and 1970 the value of agricultural production per thousand dollars of farm capital fell from 402 to 316 dollars, a decline of 23 percent. This decline was experienced in all regions of the country, except the peninsula, where the data, especially for 1950, are the least reliable of all.

This reveals a serious situation, and one which has considerable significance for the future. Furthermore, these figures measure productivity in terms of gross output; if they could be calculated in terms of net output, or value added, the picture would be worse: a decline of probably some 40 percent. Nor is this all. The census record of invested capital excludes the public investment in irrigation, in unrecovered farm credit, in research and extension, and in other items such as farm roads which directly benefit agriculture—and all these investments have been expanding. If some or all of these items were included, the deterioration in the capital output ratio would be even more severe.

If we look toward the coming years, remembering how costly it is going to be to mobilize additional resources of land and water for food production, the conclusion is inescapable that the productivity of agricultural capital must continue to fall and may well decline more rapidly than in the fifties and sixties. The probability is that by 1990 each thousand dollars of farm capital will be producing only some 200 dollars of output. If during this period production is to expand at an acceptable pace, then agricultural investment will have to increase both substantially and rapidly. And if this need is not to place too heavy a strain on public funds, ways must be found for shifting at least part of the burden to the private sector, which in turn means increasing the security and profitability of farm operations along some such lines as those suggested in the previous chapter.

Agricultural Investment

These considerations lead naturally to a brief review of agricultural investment and how it has been functioning in the circumstances of Mexico. Let us first examine the private farmers. Their investment in land acquisition has been seriously handicapped, partly by the fact that more than half the farm land of the country is not for sale, being reserved for ejidatarios and comuneros, and partly because the purchase and improvement of private farm land entails risks of violating some clause or other in the agrarian legislation and exposing oneself to expropriation. As a result, private farms rarely change hands, except the small mini-farms. Neither do farm owners (except those well within the legal limits) risk spending money on raising productivity, such as grassland improvement. And without grassland improvement, an increase in size of herd is likely to result in overstocked pastures and their deterioration.

A substantial number of cattle ranchers have financial resources (and also many crop farmers in the better irrigation districts), but they are driven to invest in real estate or in some other non-agricultural enterprise. They cannot even obtain permits to drill wells for irrigation even where the presence of groundwater is known; only the State may develop water resources.

It follows that in recent decades the history of agricultural investment is almost synonymous with that of public investment in the sector. From the time of President Cárdenas (1934–40) investment in agriculture became a major feature of government policy, and almost all of it was devoted to irrigation works. The volume of this investment increased in every presidential term, and especially in those of Alemán and Echeverría. Only in the seventies did nonirrigation investment come to represent more than 10 percent of the total, as a consequence of increasing attention to integrated rural development programs entailing not only infrastructure but also small-scale industries.

In recent years it has been fashionable in certain quarters to assert that public agricultural investment has been neglected, a thesis supported by citing the figures of this investment as a percentage of total public investment. But naturally in a country which is in process of industrialization it is inevitable that industry and related services, such as electricity supply, should claim an increasing share. Actually, measured at constant prices, public investment in the farm sector has been doubling every fifteen years, which is no small achievement. In the earlier years the returns on this investment in terms of increased output were spectacular, though subsequently the capital-output ratio worsened. This quite normal change results from the ever increasing costs of new land reclamation and irrigation works as well as from the institutional obstacles discussed in the previous chapter.

An analysis of public agricultural investment by regions suggests that most of it has gone to the prosperous ones. Yet this would be an unfair judgment: one has to remember that before public works brought irrigation to states such as Sonora and Sinaloa these were extremely poor regions. They have become prosperous as a consequence of the investment. On the other hand, the low investment in the states of, for example, Oaxaca and Guerrero is largely explained by the impossibility of irrigation, by the mountainous terrain, by the inadequate rainfall and soils of low fertility, which together make these into marginal farming areas that sooner or later will be progressively abandoned. It is neither feasible nor sensible to aim at an egalitarian geographical or per capita distribution of public investment funds.

It is quite possible to allege that even the public investment in irrigation has not been strictly speaking economic. Thus a rough calculation of total public and private investment in irrigation between the mid-thirties and the mid-seventies places the total at some 60,000 million pesos, reckoned at 1970 prices. The 1970 farm census reported the value of all crop land at around the same figure, within which total the value of the irrigated areas could hardly have exceeded 20,000 million, allowing these to be worth two-and-a-half times the nonirrigated crop land. It looks as if the total investment expenditure amounted to three times the value of the capital asset created. However, this would be an oversimplified picture. The truth is that most of the irrigation districts will have a productive life lasting several decades, even though maintenance is becoming expensive in some of them, and it is impossible to arrive at a just estimate of asset depreciation in such cases. Furthermore, for reasons already mentioned, the market value of irrigated (as of other) land has been understated and is by international standards absurdly low in relation to the value of the output it produces. Were it not for these factors, the land values would be significantly higher, and the irrigation investments of the government would appear to have been a more economic proposition.

Turning our attention from the past to the future, it is legitimate to ask what the future investment needs of the farm sector are likely to be. Some elements of this topic have already been discussed in previous chapters. For example, in trying to quantify the additional areas of land that could and should be brought into cultivation during the present decade, it was pointed out that most of these lie in zones difficult of access, largely covered with dense or medium-dense jungle, much of the land needing drainage and/or flood protection works; in short, that the capital costs per acre were going to be high by any standard.

Similarly, in regard to irrigation, since most of the large-scale new projects will be located in the south and southeast (where the topography and climate create major difficulties), the estimates in the National Water Plan have indicated unusually high per acre costs. To this must be added major expenditures in long overdue rehabilitation of existing irrigation districts.

In the livestock sector, not only will investment, mostly private, be needed in expanding and upgrading herds, but substantial sums ought to be being spent on pasture improvement—fencing, reseeding, watering points, small earth dams, etc.—to halt the present degradation and support larger numbers of cattle. Also money will be needed for dip tanks and other equipment for pest and disease control. And it is in livestock products where demand will be strongest and production expansion most needed.

It must also be assumed that the replacement of animals by machines for field work and for road transportation will continue to gather momentum, though of course the net investment will be only the difference between the cost of work animals and their equivalent in machinery and vehicles. Farm operators will invest in improving the stabling of their animals and in better housing for their own families. Government will have to continue investing in farm roads, electricity for the more isolated villages, and potable water supplies. The federal authorities as well as private enterprise should be constructing more storage space for harvested crops, the present capacity being woefully insufficient and causing considerable losses.

It would be unrealistic to attempt to quantify each and all of these items, especially since at any one time the requirements of the farm sector have to be weighed against those of other sectors of the economy. Moreover, the question arises as to how much of this investment should result from private initiatives and how much could or should come from public funds. These are matters which have to be discussed within a framework of the nation's overall development during the coming years and are reserved for the concluding chapter of the present study.

FARM CREDIT

Like any other form of loan, farm credit can fulfill one or all of three functions. It can stimulate and facilitate an increase in the borrower's productive capital, as when a farmer borrows to purchase a cow. It can provide working capital, so-called production credit used for the purchase of seeds, fertilizers, and other current inputs, and repaid when the crop is sold. It can also be used to subsidize consumption either directly to protect farmers against forced selling of their harvests, or indirectly when the farmer fails to repay his loan. In Mexico farm credit has been used overwhelmingly to finance crop production expenditure, to a significant extent for subsidizing consumption, and only in relatively small amounts for creating new productive capital.

Farm credit can be obtained from a variety of sources: private individuals, private banks and credit agencies, and public institutions. In developing countries, such as Mexico, farmers obtain loans on a substantial scale in many informal ways from individuals, be they merchants, dealers, storekeepers, lawyers, or just moneylenders. By the very nature of these transactions no data are available concerning their volume.

Private Credit Sources

Much more is known about the activities of the private banking and credit institutions operating in the farm sector. For instance the Bank of Mexico publishes figures of credit balances outstanding at the end of each year respecting both private and public institutions. From these it can be seen that the private institutions have been providing something between 30 and 40 percent of the combined total, a share which has changed little over the years. Furthermore, as measured by these year-end balances, the volume of private bank credit extended to farmers has been rising rapidly. Calculated at constant prices, it increased sixfold between 1950 and 1970 and then by a further 50 percent between 1970 and 1980.

Practically all of this credit is extended to private farmers, since the ejidatarios and most of the comuneros cannot offer their land as surety. Even so, the security of farm loans in Mexico is not as great as in other countries because of the private farmers' insecurity of tenure, the danger of invasions or of arbitrary expropriation. Hence the banking system lends substantially less to the farm sector than it would willingly do in other circumstances. On a number of interesting topics, such as who these private borrowers are, what the objectives of the loans are, and what their geographical distribution is, no information is published. We can only surmise that the major part of the lending is directed toward the commercial farmers in the major irrigation districts and to the cattle ranchers principally in the north.

Another supply of farm credit, channelled partly to the private and partly to the public institutions, is organized by the Bank of Mexico through what is known as the FIRA (Funds Established in Relation to Agriculture), a system first launched in 1954. At present it includes three separate funds: one which operates production credit, another long-term credit and the third guarantees a proportion of the loans extended by the private banks as well as reimbursing both the private banks and the State Rural Credit Bank their expenditures on extension and technical assistance.

Some two-thirds of FIRA's funds are allocated through private and the remainder through public institutions. Of the former some 30 percent is short-term and 70 percent long-term credit; but to the State Rural Credit Bank the lending is almost entirely for long-term projects, since this bank is reluctant to use its own funds for long-term loans. The great bulk of the short-term credit goes to the northern and northwestern regions, but the long-term lending is more evenly distributed geographically.

As regards FIRA's crop credit, in the early days most of it financed the cultivation of cotton and wheat, but during the seventies corn, beans, and soybeans began to assume importance. The long-term credit is devoted to the financing of beef and dairy cattle operations.

FIRA acts as intermediary for channelling to Mexican banks funds provided by the international agencies, notably the World Bank and the Inter-American Development Bank, both of which concentrate on long-term operations. A condition for the use of these funds, and for the discounting of Mexican banks' loans is that all this credit should be supervised, that is, accompanied by services of technical assistance to the borrowers, a condition which has done much to ensure positive results. Borrowers who have taken advantage of these facilities have almost without exception achieved large increases in output and in net income. Admittedly, most of these were large-scale farmers, at least in Mexican terms, operating units of 200 acres and upward; yet this demonstrates what can be done with careful supervision and holds out hopes that gradually the benefits can be extended to smaller units.

The Supply of Federal Credit

The first agricultural credit law in Mexico was promulgated in 1926 to mobilize funds for lending and to organize societies of borrowers. It established the National Agricultural Credit Bank which enjoyed an uninterrupted life until the merger of 1975 (see below). Its lending was directed mainly toward medium and large-scale private farmers. At the same time the act authorized the creation of a number of regional credit banks to service the ejidos, and these banks in turn organized several hundred credit cooperatives. This system did not prosper and was reorganized in 1935 by establishing the National Ejido Credit Bank.

During the 1940s and 1950s a number of other agencies came into existence part of whose functions was to make farm credit available. These included the National Union of Sugar Producers, the National Foreign Commerce Bank (in respect to export crops), the National Agricultural Insurance Agency, and others; but none of them ever became an important purveyor of funds. Finally in 1965, partly to establish decentralized credit operations and partly to give greater emphasis than the two big banks to long-term improvement credit, the Agricultural Bank was established, and rapidly in scale of operations became one of the big three.

In 1975 the government decided to merge these three into a single institution named the National Rural Credit Bank with the intention, only partially realized, of eliminating duplication of services and effecting economies in administration. The general scope of the activities of this bank are briefly described in the following paragraphs, on the basis of data respecting authorized credit in its first full year of operations. Although credit actually contracted usually falls short of the amount authorized in any one year, and although the scale of operations has increased since 1976, the pattern of farm activities supported and the types of borrowers have not changed significantly.

In 1976 some 80 percent of all loans authorized by this Rural Credit Bank were short-term and 14 percent long-term. About 75 percent of the total was devoted to crop loans and 19 percent to livestock programs. (Not all short-term lending is for crops and not all long-term for livestock.) The remaining loans were granted to agricultural processing plants, e.g., flour mills and slaughterhouses. Two thirds of the loan total authorized were in favor of ejidos and ejidatarios and the remainder for private farmers.

In geographical distribution the bank directs a major part of its lending to the northwest and northern regions, i.e., those in which commercial farming predominates, this being particularly the case in regard to crop credit. Livestock improvement credit assumes greater importance in the center and south, reaching nearly half the loan total in certain states. The fact that this bank contracts few loans with the cattle ranchers of the north indicates that these are accustomed to using the private banks for their financing.

Crop credit is directed overwhelmingly to ejidatarios who in 1976 received over 60 percent more per acre of crop land than did the private farmers. The amounts were particularly large in the southeast, notably among the henequen growers of Yucatán. However, calculating the average size of loan per farm operator, the private farmers registered a higher average, chiefly because they cultivate larger units.

In respect to individual crops, the loans for corn cultivation amounted to more than twice the amount devoted to any other crop; other crops of some importance were sorghums, beans, wheat, and cotton. Roughly one-third of the nation's corn acreage benefits from federal crop credit, between 40 and 45 percent of the sorghum and bean area, nearly 60 percent of the wheat and two-thirds of the cotton. In terms of values, the National Rural Credit Bank in 1976 authorized 247 pesos of crop credit for every 1,000 pesos of crop output. The ratio was above this average in the northwest and north, below average in the center and quite astronomical in the southeast. (It must be constantly borne in mind that federal credit is only one among many sources of farm finance.)

The federal credit banks have never sought to play a major role in livestock credit. Partly they claim a shortage of suitably trained extensionists to supervise this type of credit, but more importantly they have disliked tying up their funds for long periods. At the national level the livestock loans amount to a mere four percent of the capital value of the livestock population; again extremely high lending is reported from the peninsular states.

The Rural Credit Bank has continued the policy of its predecessors in devoting most of its resources to short-term lending. The great attraction of this operation to the bank is that the money is recovered (or most of it) as soon as the crop is harvested and sold. The farmer gets money for his cultivation expenses, his seeds, fertilizers, machinery hire, pesticides, etc. In some districts the bank actually purchases these items for him; also in some it takes responsibility for selling his crop, deducting from the proceeds of sale the advances it has made to him during the season. The risks are low if the extensionists do their job conscientiously, for the crops are insured compulsorily against drought, flood, and other natural hazards by the National Agricultural Insurance Agency.

On the other hand, such a policy contributes little to effecting lasting improvements in farm practices and land productivity. It can of course be argued that crop credit ensures a more widespread use of quality seeds, fertilizers, and pesticides than would otherwise occur; yet even this is doubtful since most of the credit goes to the more commercially minded farmers who would purchase these inputs anyway, while the subsistence farmers are not creditworthy and remain unserved. Moreover, once the cycle is in operation there is little change from year to year. When accounts are settled after the harvest, the farmer is in the same position as he was at the beginning of the year, and the following year the same cycle is repeated. Only through long-term credit devoted to land improvement such as terracing, draining, planting windbreaks or orchards, and to livestock expansion both in quantity and quality, can the farm's volume of production and the farmer's net income be permanently increased.

Problems and Policies

This leads to a consideration of the problems facing the federal system of farm credit in Mexico and to suggestions for modifying some of the existing policies. Although the organization of farm credit has expanded rapidly in recent decades, it was estimated that at the end of the seventies only 20 percent of the crop area benefitted from public credit facilities in spite of the emphasis that the Rural Credit Bank continues to place on crop credit.

For many years this deficiency was supposed to be due to lack of a sufficient volume of loan funds. However, even with the substantial expansion of financial resources available to the federal bank during the seventies, the coverage remains poor. It is now seen that the more important explanation lies in the lack of creditworthiness of a large portion of the farm population. For instance, in the mountain areas of Guerrero, where the only crops are corn and sesame, a subsistence farmer with five acres may obtain a cash income of only 250 dollars per year to satisfy all the needs of his family. His soils are too marginal and the rainfall too scanty to allow much improvement in output through better seeds and using fertilizer. There are probably several hundred thousand families in an analogous situation.

It has been usual in Mexico to classify farmers into three groups, from this point of view: first, those who are creditworthy and will normally repay their loans; secondly, those who with training and experience could become good credit risks; and thirdly, those who cannot become creditworthy either now or in the future. Some writers place one third, others as many as one half of all farmers in the third category.

The official banks have taken ambiguous attitudes to this problem. On some occasions they have stressed their duty to preserve the funds at their disposal and recover them for future lending operations, while at other times they have proclaimed a social duty to assist the marginal producers. At those times and in those districts where the latter view prevailed, the loan recuperation rate has naturally been unsatisfactory. Moreover, as a result of the significant expansion in the volume of lending in the mid-seventies, the proportion of marginal clients inevitably increased. And this was one reason why the Rural Credit Bank augmented its supply of credit for "living expenses" to tide small farmers over the postharvest period until their produce could be sold at better prices.

It is desirable to bring an end to this confusion of objectives. If government takes the view that a certain section of the farm population is so poor that it needs cash support, then the preferable procedure would be to set up a fund for this purpose and have it administered by the social security agencies which are experienced in the poverty problem. In any case, social responsibility for these people should not be mixed up with farm credit.

As to the intermediate group, consisting of farmers who could become economically viable, here the task is mainly one of supplying them with more efficacious technical assistance and then, step by step, with credit closely supervised by extensionists. Such a program would require close collaboration at the local level between the Rural Credit Bank and the extension staff of the Department of Agriculture.

Another perennial problem is that of organizing the borrowers. Whereas the private farmers are able to offer their land as pledge for their loans, the ejidatarios have no corresponding assets, and for this reason the credit banks have always required them to form societies with collective responsibility for their borrowings. This requirement has disadvantaged the ejidatarios, most of whom find it extremely difficult to create such societies and maintain them in being. Furthermore, they observe that in practice, once a society is formed, the credit bank proceeds to deal with each member individually in respect to loan requests and in supervising his crop cultivation.

If however, as suggested in the previous chapter, each ejidatario had a valid title to his land, usable as security for loans, this vexatious problem of the credit societies and their malfunctioning could be liquidated, and the ejidatarios with farming skill and initiative would be able to finance their operations as easily as private farmers. Of course, in the collective ejidos, the Credit Bank would continue lending to the ejido as a single unit.

A related problem is the administrative organization of the bank itself and its branches. It is widely admitted that its bureaucratic structure is over-elaborate and top-heavy, and that the costs of administering the loans are unduly high. For example, the procedures required for creating a new credit society involve months of negotiations, sometimes more than a year, by which time the participating farmers have become discouraged. In those districts where the bank assumes responsibility for purchasing the seed and fertilizer, many times these have arrived too late for use in that crop season. Many are the complaints in respect of disposal of the crop by the bank: large deductions from the price because of alleged irregularities in the quality of the produce, long delays before payment is made, and finally payment in the form of checks which small farmers, especially in remote areas, have difficulty in cashing.

The bank claims to be short of technical staff, yet it ventures into various activities far removed from the provision of credit, as for instance the operation of farm machinery centers, managing food processing plants, even administering irrigation districts. Moreover, there is much unnecessary overlapping. The FIRA programs have their extensionists, the Department of Agriculture its own network, while the Credit Bank has the most numerous army of all, since it tries to vigilate every act of the ejidatario throughout the year, and it has taken on in effect the day-to-day management of the large collective ejidos. The competition between these three services, working in parallel (and there are also others for special crops) results in farmers being given conflicting advice which saps their confidence in those who are supposed to know.

In some third world countries the extension service has been unified under the direction of a single agency; elsewhere, at least all the services relating to farmers have at the local level been brought together within a single building. If in Mexico there were less rivalry and more cooperation between the services of the Credit Bank and those of the Department of Agriculture, their respective responsibilities could be sorted out, with the bank negotiating and administering the loans as such, and the department's extensionists providing the technical advice and supervision.

A further desirable reform would be to review the interest rate policy of the Credit Bank. Considering that most of its lending is directed to the better-off farmers, it seems difficult to justify a continuation of interest rate sub-sidies. These operators can and do borrow from other sources at the higher market rates prevailing. If commercial rates were charged by the bank, it would have available a larger volume of funds and could contemplate shifting the emphasis of its lending policies from short-term production credit to long-term improvement credit. Only through such a change can the hope be realized of making the physical resources—the land, the water, and the animals—more productive. The present policy of trying to extend crop credit to ever wider acreages of semimarginal land may be a welfare service to marginal farmers, but it will not achieve expanded output. What is needed is to utilize the available finance for land improvement projects and for urgently needed livestock expansion.

AGRICULTURAL PRICES

We have examined the possession of productive capital and the availa-bility of farm credit as two preconditions for the prosperity of the farm population. But these, although essential, are not in themselves sufficient. Another set of requirements concerns the prices of the commodities which farmers buy to operate their businesses, the prices of the commodities which they produce, and the prices of the articles which they and their families purchase as consumers. An important aspect of the agricultural scene in Mexico has been the evolution and the interrelationship of these groups of prices, which have differentially affected the level of farm incomes in various regions of the country according to the farm products that characterize each area. Also it is desirable to investigate the extent to which price policies could in the coming years influence the development of production and the well-being of the farming community.

Prices of Farm Inputs

It is logical to commence the analysis with a review of the principal purchased inputs, such as fertilizers, feeds, seeds, machinery hire, etc. Since 1960 the Bank of Mexico has published annually a weighted index of the prices of these various items, and the data for selected years are reproduced in Table 9.4. Comparing the evolution of this index with that of prices in general, it can be seen how during the sixties the inputs index increased by only 10 percent whereas the general index rose by more than 40 percent. This occurred partly because the real costs of production of some of the items were falling and partly because of certain subsidies provided by the government. This differential was also maintained in the seventies.

Both factors were operative in regard to fertilizer and electricity prices which remained throughout both decades well below the average of the group. Conversely, the prices of seeds, feeds, and packaging materials rose by more than the average.

The benefits of comparatively low input prices are inevitably distributed unevenly among farmers, depending on the extent to which they purchase the

TABLE 9.4

INDICES OF PRICES OF FARM INPUTS: 1960 TO 1978

(1960 = 100)

Input	1965	1970	1975	1976	1977	1978
Seeds	124.9	131.1	319.6	361.2	447.5	519.7
Feeds	146.0	161.5	284.8	282.0	345.1	452.9
Fuel	103.6	103.8	153.0	162.0	211.9	212.2
Lubricants	109.8	111.5	173.5	181.7	213.4	212.8
Bags	110.0	153.0	323.6	399.2	545.9	614.2
Boxes	148.1	188.1	351.4	472.2	630.0	783.8
Organic Manures	113.2	125.0	139.4	154.9	194.9	239.8
Fertilizers	85.7	88.8	98.4	100.1	142.5	158.0
Insecticides	120.4	83.4	135.1	173.1	228.7	263.4
Milling & Ginning	100.0	100.0	146.7	235.9	252.7	273.9
Irrigation Dues	111.3	158.7	293.5	306.0	299.6	316.2
Electricity	95.5	96.0	72.5	74.6	137.1	139.9
General Input Index	108.0	109.7	172.5	194.9	265.0	296.0
Implicit Price Index of Gross Domestic Product	118.7	141.2	253.2	308.1	406.9	477.7

SOURCE: Banco de México, S. A., *Precios*, monthly bulletin.

different items. Thus, the fertilizer subsidy chiefly assists those who ordinarily use large quantities of fertilizer, which in practice particularly means farmers in the irrigation districts, but it does nothing for the subsistence cultivator in the mountains of Oaxaca. Similarly, the relatively rapid increase in the prices of feeds discriminated against those livestock producers who depended heavily on bought feeds and could not fully recoup their inflated costs in higher prices for the articles they were selling. Such effects tend to be more frequent and more marked, as we shall see, in a regime of persistent intervention in the price mechanism.

One item not included in the above-quoted index is the price of hired labor on farms. In a farm sector only partially modernized, like that of Mexico, the great majority of small-scale operators do not hire labor at all, but the item assumes importance on the larger commercial farms, especially on ranches and on the crop farms in the irrigation districts, and also strangely enough among ejidatarios in the poorer regions. (Our information on this topic derives from the published data for statutory minimum wages prescribed for farm workers in each of almost one hundred districts across the country, and which have already been cited in Chapter 7.)

Between the mid-fifties and 1980 the day wage for farm workers increased from an average of 5.50 pesos to one of 125.80 pesos (5.50 dollars), which was considerably more than the general increase in prices during the same period. However, if we assume that hired workers' productivity rose as fast as that of the farm population as a whole, the wage increase was compensated by an equivalent rise in labor productivity. During the second half of the seventies the statutory minimum has been raised appreciably faster than the increase in the cost of living or than any conceivable rise in productivity, which has of course squeezed the incomes of farmers depending on this type of labor. It has been mentioned that the statutory minimum is difficult to enforce in practice in rural areas, so that undoubtedly a sizeable proportion of the hired workers receives less than the minimum. But since this type of evasion is probably no more extensive now than in previous decades, the increases in wages actually paid probably parallel those in the statutory minima. Moreover, some farmers have retained the same tractor drivers or cowmen for several years and pay them more than the minimum as a reward for skills and for long service.

The Evolution of Agricultural Prices

The next step is to examine the changes in the prices received by farmers for their output. Strangely, the Mexican authorities do not publish an index of farm product prices, although all the necessary elements exist. To fill the gap in a provisional manner, we have constructed such an index simply by taking the series of value of output at current prices and dividing it in each year by the value of output calculated at constant prices. (By way of a test, the calculation was made twice, once based on the prices of 1950 and once on those of 1970, but since the two results were almost indistinguishable we have retained only the series based on 1970 prices.)

Let us look first at the price index for crop products. As can be seen in Table 9.5, this index rose by 66 percent during the 1950s, and had more than doubled by 1970. Thereafter it came under the influence of the Mexican inflation which began in 1973, and thus roughly quintupled between 1970 and 1980. In other words it rose during two decades at an average annual rate of 4.3 percent and then during the seventies at a rate of over 17 percent.

A more meaningful comparison emerges if the trend in this index is adjusted for that in the overall inflation index (i.e., the price index implicit in the gross domestic product calculation). Here, we see that in real terms crop product prices declined during the fifties by some 16 percent, remained stable during the early sixties, reached a low in 1972, and then recovered during the later seventies some 10 to 12 percent of their former losses. Yet, by the end of the decade they were still 25 percent below their real level of thirty years earlier.

Superficially this situation looks unfavorable to farm operators. Nonetheless, before arriving at a definitive judgment two other factors have to be taken into account. The first concerns the prices of farm inputs, which, as noted above, have been rising comparatively slowly. This means that, at least for farmers who used significant quantities of inputs, their income position was better than the mere prices of crop products would suggest. The second factor is the trend in the prices of the commodities which farm families buy as consumers. During the period 1970–79 this consumer price index rose from 100 to 367, while the crop prices index was rising from 100 to 395, which suggests that the purchasing power available to farm families improved slightly during this decade.

TABLE 9.5

CROP AND LIVESTOCK PRODUCTS:
PRICE INDICES, 1950 TO 1980

Year	Crop Prices Index		Livestock Prices Index		Combined Price Index
	Actual	Deflated	Actual	Deflated	Deflated
1950	100.0	100.0	100.0	100.0	100.0
1960	165.7	83.9	169.9	86.0	85.2
1965	204.8	87.3			
1966	200.4	82.2			
1967	203.5	81.0			
1968	206.2	80.3			
1969	200.8	75.2			
1970	207.9	74.5	225.7	80.9	77.5
1971	211.9	72.7			
1972	220.8	71.7			
1973	278.6	80.6			
1974	349.1	81.4			
1975	411.1	82.2	406.2	80.6	82.0
1976	511.0	83.9			
1977	612.3	76.1			
1978	758.2	80.3			
1979	821.8	72.7			
1980 (est.)	1040.0	74.3	996.0	71.0	73.1

SOURCES: Department of Agriculture, *Annual Crop Report* for value and volume of crop production; *Annual Livestock Report* for livestock products' prices. Volume of livestock products' production estimated by author (see Appendix).

NOTE: *Crop Prices Index:* current values divided by current volumes (at 1970 prices). Resultant index "actual" deflated by gross domestic product implicit price index. *Livestock Prices Index:* author's volume of output estimates valued first at current prices and then divided by same output valued at 1970 prices. Resultant "actual index deflated, as above. *Combined Price Index:* calculated by summing the crop and livestock current price totals and the 1970 price totals, and proceeding as above.

It should be remembered that the retail price index was held down by the subsidies granted to certain basic foods, e.g., in particular corn products (tortillas) and sugar. This of course benefitted the poorer farmers more than the richer ones for whom these articles figure less prominently in their expenditure (just the reverse from the input subsidies which helped mainly the modernized, commercial farms).

In livestock products the trends have been somewhat similar. During the fifties when the output of livestock products was growing more slowly than of crop products, the price index for livestock products rose at almost the same speed as the crop price index. Then, during the sixties the situation changed, with the livestock index moving somewhat faster. In the seventies the two groups kept in step.

Again, however, certain qualifications must be made. One is that we must take account of the prices of inputs, in this case especially those of feeds, which have also been rising, though slightly less rapidly than the end products. Another is the rather unquantifiable factor of improvement in the productive efficiency of farm animals. As a consequence of selective breeding, more scientific feeding, better disease control, etc., hens lay more eggs; and cattle, hogs, and poultry put on weight more rapidly—all of which reduces production costs. It seems probable that in recent years the livestock industry has been a shade more profitable than crop production, even in the rather difficult circumstances of Mexico.

To reach a view for the farm sector as a whole, we need to combine the crop and livestock branches into a single index. This price index of all agricultural products increased by 68 percent during the fifties, by a further 29 percent during the sixties and more than quadrupled during the seventies. In real terms it declined during each decade. But because of the advances in the productivity of land and animals, farmers were in reality improving their position far more than these trends indicate. What with these productivity increases and also the falling real prices of most inputs, farmers' real incomes would appear to have been rising at an average annual rate of around three percent over the thirty years from 1950 to 1980.

Price Movements of Individual Commodities

In a nation of three million farm families, an overall average does not mean much. A cotton grower in the northwest has not had the same experiences as the operator of a coffee plantation in the far south, and neither of them anything resembling that of a large-scale egg producer outside Mexico City. What matters for a farmer is not only the trend in prices but the relation between prices and costs, and this relationship varies from product to product and from region to region. In Mexico there do not exist any representative studies of production costs over a period of years, either for particular commodities or for individual types of farm, so we have to estimate profitability by another route. What is attempted here is an evaluation of farmers' intentions according to whether they increased or decreased the area under a certain crop following a change in its price. Where they extended the area under cultivation we assume that the crop was at that time profitable, at least relative to available alternatives; and vice versa.

Before proceeding to examine the inflation-adjusted prices of individual crops, it is of interest to look briefly at the long-term trends in the *current* prices of the three most basic crops in Mexico, namely corn, wheat and beans. As Figure 9.1 shows, the fifty-year period can be divided into three sub-periods: one of relative price stability up to about 1945; then one of moderately rising prices from 1945 to the early seventies and lastly the recent period of rather rapid inflation. The corn price rose in a series of steps interspersed with years of relative stability. The wheat price rose steeply during the 1940s when Mexico had become an importer, and thereafter remained fairly stable until 1972. The price of beans evinced fewer changes of rhythm but rather a persistent rise, though less steep than the other two commodities, until the early seventies.

FIGURE 9.1

TRENDS IN THE CURRENT PRICES OF CORN, WHEAT, AND BEANS: 1925 TO 1979 (PESOS PER METRIC TON)

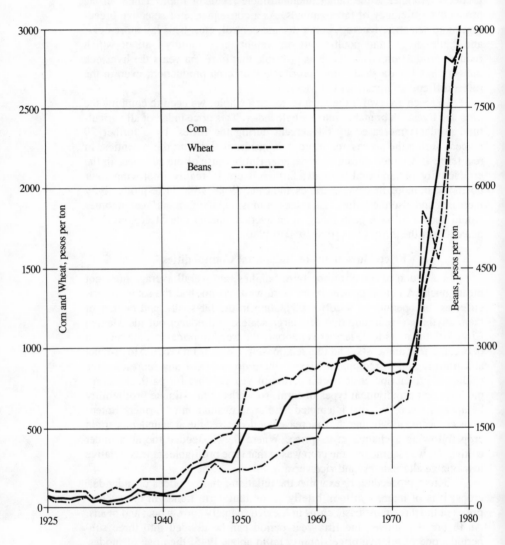

Source: Secretaría de Agricultura y Recursos Hidráulicos, Dirección General de Economía Agrícola, *Consumos Aparentes de Productos Agropecuarios, 1925–1976,* and subsequent annual reports of Dept. of Agriculture, Bureau of Agricultural Economics.

It is interesting to compare the prices of these three products. For example, the wheat price for many years remained consistently above the price of corn, but in every year since 1962 (partly due to the fixed guaranteed prices) it has remained below. One consequence of this has been a more pronounced fall in the wheat area than in the corn area. The price relationship between corn and beans is of critical importance to a large number of the poorer farmers in Mexico, and for a long time it remained steady with beans costing about twice as much per ton as corn. Suddenly, in 1974, the government fixed a new beans price, bringing them to three times the price of corn, though there was no reason to suppose that the production costs of beans had increased to any special degree. Subsequently, as a result of further adjustments, the price relative settled down to about two-and-a-half the price of corn; strangely enough, as we shall see, this price aberration had no stimulating effect on the acreage dedicated to beans. The general difficulties encountered by the government in operating a regime of fixed prices for many farm products will be considered in a later section.

It is now convenient to turn to a presentation of farm product prices in real terms (i.e., adjusted for the prevailing rates of inflation), since it is these "real" prices rather than the nominal ones which influence farmers' decision making. In the accompanying Figure 9.2 are charted, by five-year averages, the harvested area and price of each of some 24 crops which are important in Mexican farming. (It would have been preferable to cite sown areas, in order to eliminate the hazard of crop losses, but unfortunately the only data published are of areas harvested.)

Starting with corn, which occupies almost as much land as all other crops combined, we can see that its price during the seventies more or less kept in step with the rate of inflation, so that its "real" price remained remarkably stable. In previous decades its real price had been higher and had been accompanied by a steady rise in the harvested area, whereas this latter declined a little in the seventies. We may surmise that other crops were becoming more remunerative and were competing for the same land, at least in those zones where commercial farming predominates.

Wheat prices, which were extremely high in real terms in the early fifties, declined steeply but stablized in the seventies. During the first part of the decline the acreage was well maintained, largely because per acre yields were rising as a result of the improved varieties, but once this innovation had been exploited the acreage began to fall. In the other two principal grain crops, barley and rice, prices have been relatively stable in real terms for many years, but this has not apparently discouraged expansions in harvested area, in rice over several decades, and in barley during the later seventies. Mechanization, together with a gradual improvement in yields, has no doubt achieved reductions in unit costs of production.

The case of beans is somewhat peculiar. A rather stable real price in the fifties and sixties was not incompatible with a rapid rise in the harvested area, whereas in spite of much better prices in the seventies the area began to fall. We may suspect that a large proportion of the bean crop, even larger than that

FIGURE 9.2

HARVESTED AREAS AND REAL PRICES OF SELECTED CROPS: FIVE-YEAR AVERAGES, 1940–44 TO 1975–79 (1970 = 100)

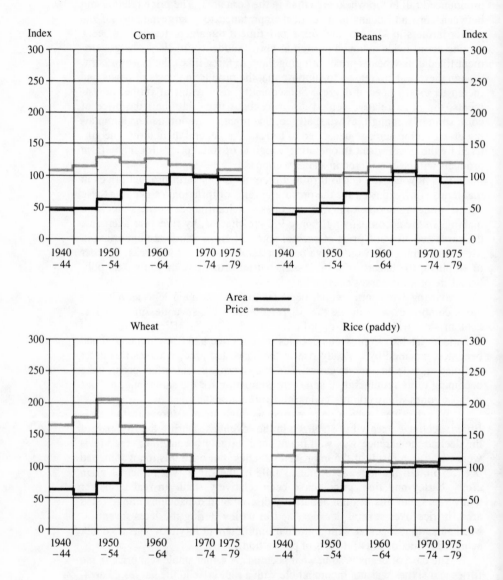

Sources: *Crop Areas*— Secretaría de Agricultura y Recursos Hidráulicos, Dirección General
(For all de Economía Agrícola, *Consumos aparentes de productos agropecuarios, 1925–*
tables *1976,* and annual reports.
in Fig. *Crop Prices*—Department of Agriculture annual reports, adjusted for inflation by
9.2) "implicit" price index of gross domestic product from annual reports of the Banco de
México, S.A.

FIGURE 9.2

HARVESTED AREAS AND REAL PRICES OF SELECTED CROPS
(continued)

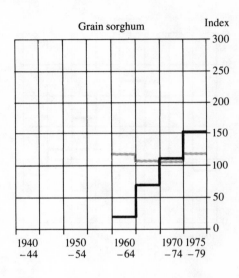

Barley

Grain sorghum

Area ———
Price ∼∼∼∼∼∼∼

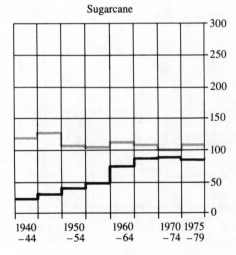

Alfalfa

Sugarcane

[219]

FIGURE 9.2

HARVESTED AREAS AND REAL PRICES OF SELECTED CROPS
(continued)

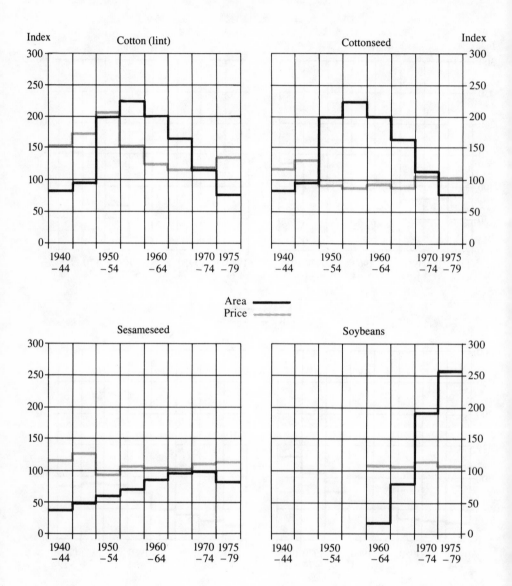

FIGURE 9.2

HARVESTED AREAS AND REAL PRICES OF SELECTED CROPS
(continued)

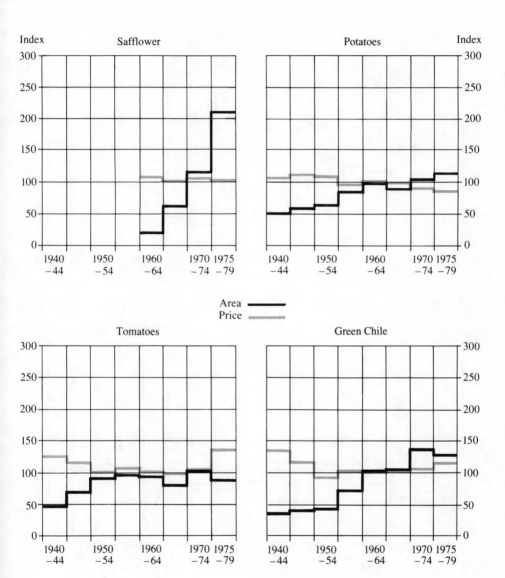

[221]

FIGURE 9.2

HARVESTED AREAS AND REAL PRICES OF SELECTED CROPS
(continued)

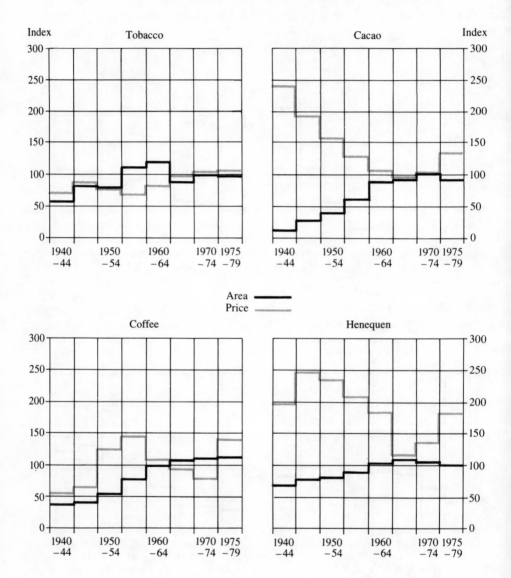

FIGURE 9.2

HARVESTED AREAS AND REAL PRICES OF SELECTED CROPS
(continued)

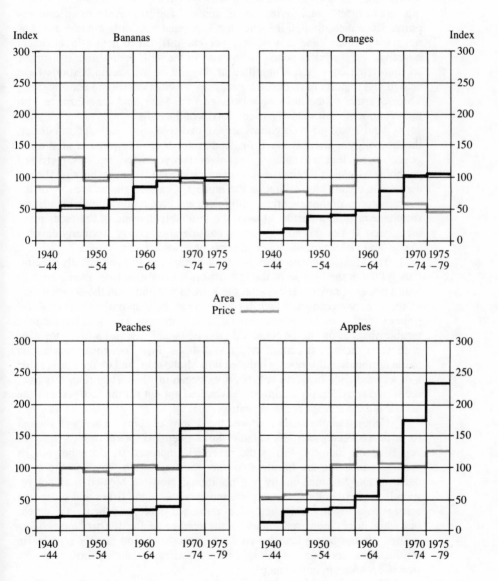

of corn, is located on subsistence farms whose operators are not especially concerned with the market price of their commodity.

Sugarcane prices, in real terms, have stayed remarkably steady for several decades, and until the mid-sixties the cane growing area was expanding fast to meet a rapidly increasing demand and to provide significant exports. Since then, though the price has kept pace with inflation, the area has not risen further, cane cutters have become difficult to obtain in sufficient quantity for the work is poorly paid, many of the mills are old and in disrepair so that much cane is left unmilled at the end of the season, exports have ceased and imports have become necessary to satisfy domestic consumption. Three-quarters of the mills now belong to the State, and cane farmers with their contracts work virtually as employees of the mills.

Alfalfa and grain sorghums are the two feed crops included in our list, though barley is used as much for feed as for beer. The price of alfalfa has never shown much dynamism, yet over the years its area has increased phenomenally, partly because of mechanization and still more because the per acre value of this crop is among the highest. Grain sorghums are one of the success stories of the seventies, with prices and production volume both rising simultaneously in an effort to meet the insatiable demand of the feeds manufacturers. The crop is now grown on considerable acreages formerly devoted to corn and yields twice as much in terms of pesos per acre.

The Mexican cotton story is one of precipitous rise and equally precipitous fall from the heights of the late fifties. It benefitted from the upsurge in world prices of raw materials after the Korean war and from the cotton policy of the U.S. government which for many years held an umbrella over world prices under which Mexico captured a share in the world market. But then the troubles began: on the physical side the pests and diseases which, together with water shortage in certain irrigation districts, wiped out cotton production in the northeast and seriously reduced it in Michoacán, and on the market side a downward slide in world cotton prices began in 1961. The price, it is true, recovered considerably during the seventies, but not till the last years of the decade did the acreage respond again.

Cottonseed obviously followed the destiny of cotton in acreage, though its price has been much less volatile. Strong demand for cottonseed from the vegetable oil industry during the seventies improved its price but not its acreage, since for the farmer it is the price of cotton which is the main determinant. Sesame, one of the traditional crops in Mexico, has not responded in textbook manner to price movements. In the fifties and sixties the acreage rose steadily while real prices remained stable, and then in the seventies while prices increased somewhat the acreage declined. This crop is being pushed off the better land by competing products, and chiefly remains in semimarginal areas of low rainfall. Copra is another traditional oil crop now slowly declining in importance.

The two new star performers have been soybeans and safflower. Making their appearance first in the late fifties, they expanded steadily, and in the seventies dramatically, to become the principal sources of vegetable oil. In both crops the acreage more than doubled during the seventies, though this caused no weakening in the price of either one. Both have come to occupy large areas in the irrigation districts where they can be grown as a second crop on land already harvested earlier in the year, which greatly reduces their production costs.

In order to illustrate some trends among the large number of crops in the vegetables group, we have chosen three distinctive examples: potatoes, green chiles, and tomatoes. Potatoes are one of the products where, in spite of no special stimulus from the side of price, the area cultivated has continued to expand, due largely to the large increases obtained in per-acre yields. However, it seems possible that the yield increases are too big to be true and result rather from corrections in the methods of crop reporting. This appears even more the case in respect to green chiles where the alleged expansion carries the implication of a doubling of per capita consumption in fifteen years. The Mexicans are traditionally big consumers of chiles of all kinds, but a doubling of consumption from an already high level would have been something noted and commented upon. And in spite of this, the price has risen faster than inflation!

Tomatoes differ from potatoes and chiles in that they depend almost entirely on irrigation and about one-third of the output is exported; indeed, this proportion is really higher because of the quantities of tomato products exported. The price, after a long period of stability in real terms, shot up surprisingly in the second half of the seventies, provoking a further sharp increase in production. That this latter increase took the form of a tripling of yield per acre in fifteen years is difficult to believe.

Tobacco has been one of the crops with the greatest stability in both area and real price from decade to decade, partly due to the regime of administrative controls imposed on both growers and manufacturers by government. National consumption has increased slowly, with per capita consumption actually falling, and has been met entirely by increases in per-acre yields.

The trends in coffee prices and cultivation have been peculiar. While until the late fifties the considerable rise in Mexican (real) coffee prices brought about a predictable expansion in the planted and harvested area, for the next fifteen years prices were falling and yet the area continued to expand. Coffee trees of course take a while to come into bearing, so one would expect a time-lag before the depressed prices would affect the area, but fifteen years should be more than sufficient. A partial explanation is that the majority of coffee growers are small-scale and poor, living in districts which offer no alternative crop; the opportunity cost of their land and labor must be near zero. In any event, since coffee is a crop which over the long term has registered no improvement in yield per acre, the growers certainly suffered a decline in incomes—until the 1975 upsurge in world prices came to their rescue. The subsequent high prices do not seem to have stimulated much new planting.

Cacao prices declined in real terms during every decade after the Second World War until they bounced back in the seventies, yet the cultivated area expanded persistently. This anomaly may be partly explained, as in the case of wheat, by the sustained improvement in yields. Inasmuch as some 60 percent of the output is exported, the price will continue to be influenced by the traditionally violent fluctuations in world markets.

Henequen has faced quite a different situation. It is a hard fiber which over the years has experienced greater and greater difficulty in competing with sisal from Africa and South America. Although prices declined steeply from the early fifties until 1971, the harvested area continued its slow expansion, and this happened not because of any improvement in yields or decline in growers' production costs (which did not occur), but mainly because the growers were increasing in number, and on the extremely shallow limestone soils of Yucatán there was and is no alternative crop. From 1972 onward the government raised the guaranteed price several times and supported it with a massive subsidy, mainly as a social welfare rescue operation. Henequen is a crop requiring seven years from planting to first harvesting, and for some years the new plantings were diminishing. Even in the late seventies with so much price incentive, the new plantings barely sufficed to replace the older plants going out of production.

Lastly we come to the fruit crops, of which Mexico boasts an enormous variety from temperate to tropical, from apples and cherries to avocados and mangoes, but until the seventies none was grown on a large scale except oranges and bananas. Four fruits have been selected for presentation in the graphs. The banana plantations showed for many years a tendency to expand without any stimulus from the price side; but the seventies ushered in a sharp drop in real prices which indeed began to affect the acreage harvested. Another factor was the damage caused by Panama disease on the roatán variety.

Oranges for many years showed, as might be expected, a tendency for the harvested area to expand some five years after a rise in prices, but recently this logical sequence has been broken. Real prices declined from the mid-sixties, and very sharply during the seventies, yet the area attained its highest level ever. Per-acre yields have not improved, so it would seem that the growers are worse off than they were a decade ago. A similar price fall has been suffered by lemons (not shown on the graph), but with the astounding response that the area doubled during the seventies.

Apples and peaches, on the other hand, with firm (real) prices during the seventies have evinced the predictable response of increases in their areas which up till this writing have not led to glutted markets and price reactions. The same phenomenon has been experienced with avocados, mangoes, and grapes, all doubling their output during the seventies and not showing any price weakening, except in the case of grapes which are grown chiefly for wine making and distillation. It would seem that the Mexican public, for decades a modest consumer of fruits, has begun to acquire a fruit-eating habit.

The foregoing analysis of the trends in prices and harvested areas reveals how complex are the relationships between the two and what a wide variety of factors have to be cited in order to explain what happened. Some of the factors are indeed common to many crops, others to only a few. Neither do those which are determinant in one decade necessarily maintain their significance in another. The reasoning process has to be adjusted in the light of changed circumstances. Nevertheless, from this historical review a few characteristics stand out as worthy of mention.

First, over a long period lasting until the late sixties, the area of nearly all the crops studied tended to expand, in some cases rapidly and in others more slowly, despite the fact that real prices of these commodities were either (a) more or less stable, a characteristic of the majority, or (b) were falling, as for example in wheat, coffee, cacao, and henequen. The persistence of this phenomenon has two explanations. First, in a number of products the costs of production per ton were falling which enabled farmers to accept lower prices without hardship. Secondly, where production costs were not falling, or not sufficiently, farmers were obliged to accept lower real incomes, especially in regions where they could not escape into other more lucrative crops.

From the beginning of the seventies certain new tendencies became apparent. Apart from a few notable exceptions, the real prices of most farm products have improved, in some instances substantially—a consequence in some products of governmental price fixing and in others of a modified pattern of consumer demand. Yet notwithstanding that fact, the areas of many important crops, far from augmenting in response have actually declined, for example, corn, beans, wheat, sugarcane, cotton, tobacco. It is probably not fanciful to trace this change to a gradual change in the opportunity cost of farm work. The pressure of population on the land, though still acute in particular regions, is beginning to diminish. Even if whole families do not move away, at least some family members find part-time employment in nonfarm occupations. As a consequence, many farmers who were under pressure to expand their production in face of declining prices merely to maintain their ever-growing families, today can supplement their family income from other sources, and hence it is for them adequate if their farm output is maintained at a constant or slightly declining level unless product prices turn particularly attractive. If this be so, at least in some degree, we must expect that during the eighties the price stimuli for any desired level of production may have to be stronger than in the past.

This leads naturally to a consideration of the government's price policies in regard to agricultural commodities, but before tackling this theme a few words are needed about trends in the prices of livestock products and their influence on the levels of production. This is a more hazardous exercise than in the case of crops, for, although annual price statistics are published for milk, eggs, various types of meat, and other items, estimates of production were formerly confined to the decennial censuses and have become annual only since 1970 under the auspices of the Department of Agriculture—a series which is as yet far from being credible. As explained earlier, because the censuses so severely underestimated the volume of production, the author developed independent estimates of output which are used in the accompanying graphs (together with those of the department from 1970 onwards).

In the case of beef, both real prices and the volume of production rose steadily until the end of the sixties, but thereafter prices stagnated. In spite of

FIGURE 9.3

ESTIMATED QUANTUM OF OUTPUT AND REAL PRICES
OF SELECTED LIVESTOCK PRODUCTS: 1950–54 to 1975–79

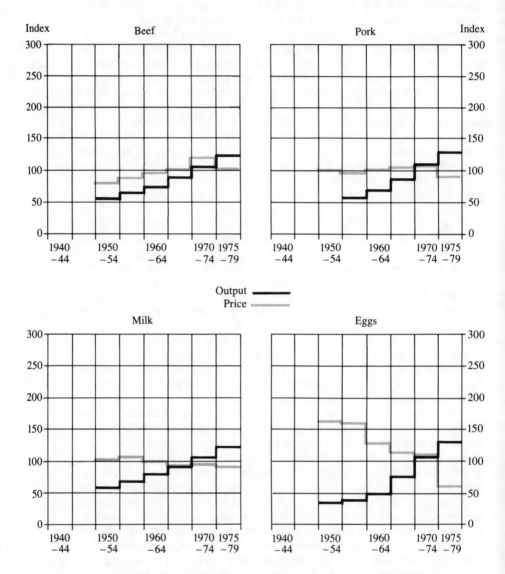

Sources: *Livestock Prices*—Department of Agriculture Annual Reports, adjusted for inflation
  by "implicit" price index of gross domestic product from annual reports of the Banco
  de México, S. A.
  Output of Livestock Products—author's estimates

this the output of beef continued to rise rapidly throughout the seventies, according to the official figures. But these latter, when compared with the data on cattle numbers from the same source, show an increase in beef output per animal of around 26 percent during the decade, which is scarcely plausible. Altogether it is impossible to draw any firm conclusions.

A similar problem presents itself in respect to pork production. Real prices seem to have remained stable for a long period, but the output of hog products registers an uninterrupted increase; and during the seventies, as in the case of beef, output was based on a presumed rise in the productivity of hogs, in this case amounting to over 50 percent. In Mexico, whereas most beef cattle are grass-fed up to slaughter, hog production depends on feed concentrates, the prices of which have been moving upward, thus putting a squeeze on producers' profits.

Milk prices and production followed trends similar to those of hog products up to the end of the sixties, but the act of shifting to the department's estimates for the seventies would show an increase in output of over 40 percent paralleled by virtually no increase in real price. Whatever the official statistics might say, production certainly declined at the end of the decade. In 1979 (and still more in 1980) milk became scarce in the larger cities, and consumers stood in line for deliveries which might not arrive. The dairy farmers' association reported widespread slaughter of cows because production had become uneconomic at the prevailing fixed price.

Eggs provide a classic example of the effect of improved technology on production costs. From the fifties to the seventies the real price of eggs declined steeply, yet production by the new techniques remained profitable and output expanded rapidly. According to our estimate, this attained a rate of some two percent per laying hen per year. However, egg yields on commercial units are already high and such a rate of improvement is unlikely to have been sustained throughout the seventies, nor will it be in the eighties.Probably the population of poultry for egg production will continue to increase, though high feed costs will sooner or later slow its growth.

Unfortunately, the livestock products data are too uncertain to permit drawing any meaningful conclusions. If the official annual estimates are discarded for lack of credibility, we are left with our arbitrary assumptions as to rates of growth of productivity, which are unable to take into account many of the important variables, such as changes in feeding costs and modifications in the ceiling prices imposed by the government. A complete overhaul of the data reporting service in the livestock sector is urgently required.

Price Controls and Price Policies

In Mexico, as in other countries, government intervention in the pricing of farm products has become widespread. For a number of crops it takes the form of establishing guaranteed prices with or without some State purchasing activity. In other commodities it takes the form of setting ceiling prices at wholesale and/or retail level. Where the fixed retail price is so low that distribution costs cannot be covered, the government provides a subsidy, notably for the so-called "basic" products like tortillas, sugar, rice, and beans. A few products are subject to "administered prices," in that they are entirely purchased by semigovernmental agencies, for example tobacco and henequen.

The system of guaranteed prices was first elaborated in Mexico in the fifties for corn, wheat, and beans, and was subsequently extended to other crops until at the end of the seventies it covered thirteen products. In the early period, when national inflation was 5 percent or less, these prices were set and retained without change for a sequence of years. For instance, the guaranteed price of corn stayed at 940 pesos a ton from 1963 to 1974, that of wheat at 913 pesos from 1955 to 1966, that of beans at 1,750 pesos from 1961 to 1973. During the seventies, with inflation at 15 to 25 percent instead of the 3 to 5 percent of earlier years, the guaranteed prices were raised annually, but in few cases did the rise fully compensate for the rate of inflation.

Furthermore, the price rises appeared arbitrary and resulted in distortions of the price relationships hitherto established by the market. For instance the price of beans was tripled in a single year, while that of rice was raised by only 10 percent over five years. Certain field studies of costs were made but these were neither geographically representative nor methodologically sound.

North American readers are familiar with price support programs, so it is unnecessary to list here their general advantages and disadvantages. In Mexico, however, there have been special considerations peculiar to that country. Thus, one of the arguments used to justify the program was that it would protect small-scale farmers who were vulnerable to exploitation by unscrupulous middlemen. The government has a large and heavily subsidized agency, CONASUPO, with responsibility for purchasing farm products, buying needed food imports from abroad, and running a chain of grocery stores for poor people. In theory if private buyers tried to extortionate, the farmers could at least obtain the guaranteed price from CONASUPO. In practice farmers found the CONASUPO conditions of purchase too exacting (large deductions for quality defects, etc.) and therefore preferred traditional channels of sale, and in many parts of the country CONASUPO was not represented at all. Even in the crucial post-harvest buyers' market the producers were supported more by the Credit Bank than by the official purchasing agency.

It is true that when the price support program was first started, it appeared as if the farm sector were heading for a period of surpluses, and indeed in the sixties the free market price of a particular product was usually below its guaranteed price. But by the beginning of the seventies, and thereafter, the reverse was the case; market prices were usually higher than the guaranteed prices. Farmers obtained better prices from private merchants, which is another reason why CONASUPO has seldom bought more than ten percent of any one crop.

Another difficulty has been the slowness of the governmental machine in responding to changed situations. When a group of producers claims a price increase because of higher production costs, an elaborate dossier of supporting data has to be assembled, then in the government price-fixing bureau an endless number of memoranda have to be written, consultations held with the

Department of Agriculture and other interested agencies, and proposed modifications passed upward for approval, so that often a year elapses between the petition and the decision, by which time costs will have risen further and the approved adjustment is inadequate.

For example, the sale of milk at retail level is subject to a ceiling price. Faced with rapidly rising costs during the inflation of the seventies, the dairy farmers and milk processors on several occasions claimed a price revision, and each time the deliberative process took eight to nine months. By then production costs had again risen, so that the operators were never liberated from their income squeeze, with a result that fresh milk has been periodically scarce because a number of dairy farmers went out of business.

A further problem is the inability of bureaucrats to judge the appropriate price relationship between two commodities. For instance, in Mexico the coconut oil content of a ton of copra is 59 percent, and the relative prices of the two products in a free market reflect this fact. Yet at one moment the authorities decreed a substantial rise in the fixed price of copra while maintaining that of coconut oil at its old level; consequently for nearly a year the oil mills lost money and some went bankrupt. In some countries the bureaucracy learns from its mistakes and gradually improves its performance; but in Mexico the staff of the agencies changes every presidential term, and sometimes more frequently, so at any one time hardly any of the administrators have previous experience.

Another disquieting feature is the astonishing and often unpredictable size of individual price changes. It is not unusual for the price of a certain commodity to be increased by 20 or 25 percent at a single adjustment, not to mention again the tripling of the bean price. A major reason for this is the lack of crop storage up and down the country. Capacity is so inadequate that after a large harvest it may take months in some regions to bring all the crop under cover. Thereupon the authorities refuse a price increase for the following year (thus reducing the price in real terms) in the fear of facing a second bumper crop. The farmers dutifully reduce their sowings, the weather turns bad, the harvest fails dramatically and the government is obliged to import that particular product. Living from hand to mouth encourages dislocations in supplies, but this will persist until funds are devoted to constructing a much larger volume of storage accommodation in line with the country's needs.

A few words are in order at this point concerning the massive food subsidies paid out by the government, though these affect farmers only indirectly. For many years now it has been considered obligatory to maintain the retail prices of certain commodities, notably corn products (tortillas), sugar, and a few other products at artificially low levels. These were justified originally on the ground that distribution costs were phenomenally high and that the poor must be safeguarded. With inflation running at 20 percent or more and with retail price adjustments being made only every two or three years, and then very small ones, the prices of these products fall more and more out of line and the subsidies become ever more massive.

Consider as an example the case of sugar, the Mexican retail price being one of the lowest in the world, and the per capita consumption one of the highest. The subsidy in 1979 exceeded 600 million dollars. What have been the consequences? A population that used to be renowned for its good teeth now suffers from a high incidence of dental caries, while diabetes has become widespread. The freezing of the retail price has made it impossible, even with the subsidies, to raise the cane price sufficiently to stimulate the needed expansion of production; so that what were formerly significant exports have ceased, and Mexico has become a sugar importer. Furthermore, the sugar mills have had their operating margins squeezed so severely that they have been unable to renovate or even adequately maintain their equipment, thus each year part of the harvest is left unmilled. All this and more can be debited to the government's psychosis in regard to the retail price of a commodity which has no nutritional value and which consumed in such large quantities is injurious to health.*

After so many paragraphs of criticism of price policies and supports, it is only right to formulate a certain number of positive proposals for reform. Let us consider first the input subsidies which hold down, or moderate, the prices of fertilizers, irrigation water, tractor fuel, electricity, and sundry smaller items, and which as we have pointed out chiefly benefit the larger commercial farmers in the more privileged regions. These subsidies are income-regressive in their effect: though neutral in respect to the poorer farmers, they help make the rich ones richer. It has been suggested that these subsidies should be retained only for certain classes of farms (or incomes) and certain regions—the poorer ones. But the administration of such an arrangement would be formidable, and under Mexican conditions would invite too much evasion and corruption. A more sensible solution would be to phase them out over a period of three or four years. After all, they accrue mainly to a group sufficiently prosperous to manage without them, while the poor farmers can be helped more effectively in other ways (see below).

A number of people argue that the price support program should be used as an instrument for income redistribution in the farm sector, an argument which enjoyed acceptance in the United States in the fifties and sixties. But unless a country is to be swamped with unsaleable surpluses a program oriented to this objective needs to include a system of quotas either by acreage or by volume of output. Also a government has to have available a large amount of storage capacity, either its own or rented, for stockpiling purposes to avoid reducing the supports after an abundant harvest. The U.S.A. could fulfill these two preconditions, but Mexico has neither the administrative expertise and conscientiousness to operate quotas nor the funds to operate large-scale storage programs.

Moreover, the use of price supports for maintenance of income may work fairly well in a country where the gap is not too great between the higher

---

*In the summer of 1980 the retail price of sugar was suddenly more than doubled. This has substantially reduced, but not eliminated the subsidy. But sugarcane growers are not expected to produce more, the cane price still being unattractive. Nor could the dilapidated mills crush larger quantities of cane. Mexico expects to import 900,000 tons of sugar in 1981, almost one quarter of the national requirement.

income and lower income farmers, but in a developing country where half the farm population operates on a subsistence basis, redistributive objectives cannot be attained in this way. A corn price which is barely adequate for a poor peasant in Oaxaca spells riches to an irrigation farmer in Guanajuato.

Should we therefore advocate the alternative of dismantling entirely the present regime of guaranteed prices? In practice it has always followed the market price rather than influencing it. The theoretical argument for a free market is that it encourages a gradual concentration of production into the hands of the most efficient producers. But this reasoning is valid only where there is also a free market in land, which is not the case in Mexico. Here, unless the agrarian reform institutions were modified, there would be no turnover in the ownership of farms.

It is consequently worth considering whether it might be advisable to introduce, at least for a few key commodities, a system of differential prices designed to reduce the income-regressive effects of the single price regime. That is to say, there would be a series of three or four prices for, say, corn—each price applying to a certain category of producer. Such categories should ideally be established on the basis of costs of production, but the requisite data would be unobtainable. A practicable alternative, operated at one time for wheat in France, would be to differentiate the price according to the quantity sold; e.g., the highest price would be obtained by the producer selling less than five tons per year, a rather lower one for producers selling five to ten tons, and so on down to the lowest price applicable to the largest producers.

Such a system might indeed give rise to a certain amount of evasion: a large producer might arrange for a small one to sell part of his crop and thus obtain a higher price. But the small producer would not cooperate unless he were given a share in the profit, and this in itself would be an act of income redistribution—the object of the proposal. Also, for reasons of transportation costs, such manipulations would be limited to zones where large and small farms were physically close to one another, and in Mexico there are not many of these.

In one other direction the system of guaranteed prices, if retained, could be given a positive orientation. In an earlier chapter we commented that among the obstacles to livestock expansion were the high prices of feeds in Mexico, but also noted that in the eighties an increasing volume of feeds imports will become necessary. Mexican grain and oilseed prices could be kept relatively high to encourage production while an increasing proportion of the supply would be imported at prevailing (lower) world prices. Thus the raw material of the feed concentrate manufacturers would be cheaper than if all of it were home-grown, and the prices to the feedlots and poultry producers could be reduced. And this would be reflected in the prices of meat, milk, and eggs.

There remains the question of the subsidies to retail food prices. If the low price, subsidized commodities were reserved exclusively for purchase by the low income groups, there might be indeed a justification for retaining the system; but they are not. Even CONASUPO, which originally was supposed to establish its outlets in the poor quarters of cities, now has many of its stores in middle-class suburbs. In any food subsidy thus generalized, there occurs an inevitable leakage. The subsidized sugar bought by the poorest peasant is bought at the same price by the millionaire, and the millionaire buys larger quantities; as in the case of farm inputs, a significant proportion of these food subsidies benefits the wrong people.

For directly helping poor people a more efficacious technique might be that of food stamps in one or other of their various alternative forms. In countries where such a program has operated for many years, its administration may have become open to some abuse; but in one where most of the poor are really poor and not just malingering, there is a strong case for such a program, at least during a limited number of years.

If such a program were adopted, not only could the hideously expensive food subsidies be discontinued but also the clumsy attempts at maintaining ceiling prices on certain commodities. These ceilings, always adjusted too little and too late, progressively discourage the expansion of production. They force manufacturers to adulterate or otherwise reduce the quality of their merchandise. They provoke scarcity rather than stimulate plenty. They impede adaptations to changes in consumer demand. They create the illusion that certain commodities, indeed quite a considerable number, can be had for less than their cost of production. They encourage the public to forget that "there is no such thing as a free lunch."

Improvements in the functioning of the price system both for farmers and for consumers, can occur only slowly as the overall modernization of the country proceeds. The rationalization of marketing, a topic outside the scope of this study, can come about only after chronic unemployment has been substantially reduced, else too many people will be continually entering the distribution chain. The inefficiency and corruption in the administration of price policies can be eliminated to the extent that education is generalized and professional training becomes more widespread. Twenty years hence there will be few districts, even remote ones, where farmers are not well acquainted with the market prices of their produce and where consequently they will still be susceptible to being exploited. The gradual social and economic integration of the Republic will of itself help farmers to secure better rewards for their efforts in feeding the nation.

10. On- and Off-Farm Employment

Where farming is not yet modernized nor fully commercialized, most agricultural work is seasonal and therefore part-time in character. It is, for instance, meaningless to state that there are 5.1 million persons employed in agriculture in Mexico, or to contradict this with an assertion that the figure should be 6.1 or 7.1 million. What we have to visualize, as noted earlier, is a farm labor force of quite different sizes at different times of year—a gigantic sponge that expands to around 10 million persons during the weeks of harvest and contracts to perhaps less than three million in the dead season.

We also have to visualize wide differences in the number of days per year worked by individuals. A tractor driver on a large commercial farm in an irrigation district may be a permanent employee working 300 days in the year, whereas a subsistence farmer growing corn on five acres in Guerrero may actually work only 60 days. The amount of work generated depends on the type of farm, on its size and location.

THE STATE OF FARM EMPLOYMENT

It is commonly said, and with much truth, that in Mexico as in other developing countries the rural areas are overpopulated and suffer from chronic underemployment. Yet distinctions have to be made. Thus it is doubtful whether many of the big commercial farms are overstaffed, since their operators hire no more workers than are needed. At the other extreme a mini-farm of five acres whose operator has, say, six children, some of them already adult and all living at home, certainly cannot employ them all. And this situation becomes far worse where the government has decided to transform a few hundred mini-parcelas into a collective ejido, equipping it with a full range of machinery. In these cases, of which there are many, the underemployment becomes a nightmare.

But of course the occupiers of mini-farms and their families do not sit around idling all the year. Many family members migrate as seasonal workers for cotton picking, cane cutting, coffee harvesting, and return home when the season for these activities terminates. Many engage in non-farm occupations either part-time or full-time: as bricklayers, gardeners, domestic servants in nearby cities, as storekeepers or store assistants in the village, as freight carriers if they can contrive to hire or own a truck. This is a familiar situation in many countries. Thus in the Federal Republic of Germany in the early seventies, among all farms smaller than 12.5 acres 83 percent of the operators obtained more than half their income from non-farm sources. In Mexico, opportunities for supplementary income from non-farm employment are relatively high in the rural areas near the larger towns, but almost nonexistent in the more remote districts.

We should also recall the finding in Chapter 6 that in the poorer states the ejidatarios devote most of their production expenditures not to seeds and fertilizers but to hiring other people to come and work for them. Sometimes this may indeed be because they themselves have more remunerative jobs, but often it amounts to a matter of prestige to be able to afford hired labor while they sit gossiping in the village square.

For all these reasons it would be imprudent to dogmatize as to the volume of employment really generated on farms, or about the extent of underemployment. The most underemployed peasant may work 17 hours a day during his harvest. An operator cultivating several crops maturing at different seasons, or one maintaining cows, hogs, and poultry in his yard, has employment virtually all the year round. A mechanized farm may have its employment requirements more evenly distributed through the year than a subsistence farm, yet the former may not be offering more man-hours of work per acre per year.

LABOR PRODUCTIVITY

In agriculture, as in other sectors, a distinction has to be made between mere employment and the generation of productive employment. Because Mexico has such a wide range of farm types, stretching from the one-crop subsistence unit to the diversified and irrigated commercial farms, the contrasts in labor productivity are naturally great. Difficulties are encountered when we try to arrive at reasonably credible calculations, partly because of the problem of assigning a partially mobile farm population to geographic regions, and partly because we need to include livestock products as well as crops and our estimates for the former are particularly shaky since they require assumptions as to regional differences in livestock productivity.

Using a methodology described in the Appendix and the data from the last farm census (the *relative* situation of the regions has not changed appreciably since), the value of gross farm output per person occupied is presented in Table 10.1. The peninsula registered the lowest labor productivity at 5,453 pesos and the northwest the highest with 25,340. When reckoned by individual states the performance ranged from 3,683 pesos per person in Oaxaca to an average of 54,010 in Sonora.

It must, however, be remembered that these are figures of *gross* not net output per person. Inasmuch as the expenditure on inputs is far greater on the commercial farms of the northwest, the real contrast in productivity will be considerably less, but would still remain substantial. This is also because the output data in the census were seriously underreported, and this under-

TABLE 10.1

**VALUE OF FARM OUTPUT PER PERSON OCCUPIED
IN FARMING, BY REGIONS: 1950 AND 1970**

(in 1970 pesos)

	1950	1970	Increase (1950 = 100)
Northwest	12,665	25,340	200
North	10,297	19,815	192
Northeast	9,906	20,278	205
North-center	4,930	10,609	215
West-center	4,073	8,593	211
Center	3,214	5,982	186
Gulf-south	4,071	7,478	184
Peninsula	4,387	5,453	124
Mexico	5,317	10,520	198

SOURCES: Production data from Table 1.5; area data from Table 3.1. For farm population data, see Appendix.

reporting would have occurred mainly on the larger farms, for fiscal and political reasons.

The performance of the ejidatarios may also be distinguished separately.[1] Again there emerge similarly striking contrasts between the regions, from around 1,600 pesos per ejidatario in the center region and the peninsula up to nearly 11,000 pesos in the northwest. It may be guessed that for the country as a whole the labor productivity on private farms was of the order of three times that registered among ejidatarios and comuneros.

To these differences in productivity livestock farming has made an important contribution. Of the eight states with per capita productivity exceeding 20,000 pesos in 1970, six had livestock predominating over crop production; and of the five with productivity below 5,000 pesos four had predominant crop production. This has been chiefly a response to certain ecological situations; for instance in the north where there is little irrigation only extensive ranching is feasible, and that requires minimal manpower.

Productivity is low in some of the central and southern states where population pressure is severe, the quality of the soils is marginal, and chiefly low-value crops such as corn and beans are grown. It is in these states where livestock expansion could improve farmers' incomes; and since these zones lack the extensive grazing lands of the north, the development would have to be of the intensive type with reliance largely on purchased feeds.

Between 1950 and 1970 labor productivity in real terms roughly doubled, a rate of growth of 3.5 percent per year, which can be considered an excellent performance for the farm sector of any developing country. The increase was somewhat faster in crop than in livestock production, this being

---

[1] For estimates according to class of farmer we are obliged to use only the census data since we cannot allocate on any sure basis the unreported livestock output between ejidatarios and private farmers.

especially the case in the northwest and west-central regions. However, the reverse was true of the northeast, and the Gulf and south regions. What seems to have occurred was a process of catching up by some of the states formerly very poor. Thus in regard to the more reliable data of crop output per capita, and using 1970 prices, there were in 1950 28 states out of 32 with productivity below 4,000 pesos, but by 1970 this figure had dropped to 15 states. In other words, 13 formerly low productivity states had passed to the higher group.

This is an evolution very similar to that which we have already observed in respect to rural living standards in Chapter 7. Nor should that be surprising, since in a market economy living standards cannot be widely divorced from productivity in the long run. Although no new census figures are available at this writing, visual evidence indicates that the same process is continuing, so that some of the states where poverty was still bad in 1970 have accomplished some catching up.

OUTLOOK FOR LABOR PRODUCTIVITY

The improvement of farm incomes depends basically on two factors: the man/land ratio and the yields that can be obtained from that land. Mexico's agrarian reform was in the first instance an exercise in making the man/land ratio more egalitarian; but when in subsequent years the demographic explosion became a menace, it also became an exercise in sharing the amount of work available. Each ejidatario was given on average some 25 acres, which in the next generation became 10 acres or even 5 if he had several sons. On 5 or even 10 acres a large family represents overmanning and productivity is low, however much technical know-how is brought to bear. But most of these sons, had they migrated to the cities, would have found no work at all, though those remaining behind would have had more land per family and higher labor productivity. In other words, the amount of work available over the country as a whole was more equally shared, at the price of retarding the growth in productivity of certain groups of people.

Gradually the situation began to change. The expansion of the economy in the sixties and seventies provided more job opportunities but at the same time augmented the disparity between rural and urban incomes. From egalitarianism in capital assets (land) the emphasis shifted to a plea for greater egalitarianism in incomes. It was observed that output per acre was extremely low on mini-parcelas, and those people inclining toward a socialist ideology believed that by reorganizing the ejidatarios into collectives, mechanized and modernized, incomes of the members would materially improve. During the presidency of Luis Echeverría quite a number of collective ejidos were established, but the results were disappointing, as could have been predicted. The administration of any large farm is a daunting task, and more especially in Mexico where there was nobody trained in this type of management and where most of the ejidatarios were passively resisting the experiment foisted on them.

In the worst of the collectivized ejidos the collective income fell below what the individual ejidatarios had obtained when on their own; in the best it marginally improved on their previous performance, but at the price of leaving many of them under or unemployed, because of the ambitious mechanization. The managers were not permitted to discharge their surplus personnel, and the net income had to be shared among those who worked and those who did not—not exactly an incentive to work hard. With but few exceptions these collectives were slovenly cultivated and their animals carelessly tended, because they failed to mobilize the individuals' instinctive interest in self-advancement.

These disappointments should not be interpreted to mean that the outlook is hopeless, merely that other avenues should be explored. In some quarters there has been much talk of crop diversification as a promising solution: a shift from the corn/beans syndrome to high value crops such as vegetables and fruit. With such crops a large volume of sales can be obtained from a small acreage and almost year-round employment provided. Although it is true that the domestic demand for certain vegetables and fruits is rising rapidly, and there are some export possibilities, all this has its limits. Already the market for a few of these commodities is showing signs of saturation. These intensive branches of production can indeed offer opportunities for a restricted number of farmers located in favorable climatic zones, but can hardly constitute a solution for two million poor farm families.

It is often more encouraging to examine some of the real-life experiments being tried out in particular districts where ingenious individuals work out ways of overcoming local problems. One such example can be found in the desert area of northern Coahuila, where the rainfall varies between 150 and 300 mm, where the pastures are degraded, and where almost the sole activity of the ejidatarios is raising goats. Here a modest program of pasture improvement was introduced (the grazing being collective but not the goats), with wells, water tanks, fencing, and where feasible small earth dams. This was accompanied by a campaign to persuade and help the farmers to purchase goats of improved breeds yielding more milk, as well as arrangements for processing into goat cheese and providing market outlets. The cost was small but the income improvement substantial.

Another example: some agronomists in the state of México discovered that fodder oats and various grasses for feed could be successfully cultivated in forest clearings at altitudes of 8 to 10,000 feet. These crops could be cut for hay or silage or grazed by animals in fenced paddocks. The climatic and ecological conditions of this region are replicated in many of the mountain areas of the central states where the adoption of this innovation could significantly increase the carrying capacity of the land and the livestock output.

A quite different innovation proved successful in the central zone of the state of Veracruz, where normally only one crop of corn is raised per year, with beans between the rows. With irrigation of the simplest kind—a well and one small pump—and with levelling of his land, a farmer could contrive to grow three crops of corn in a twelve-month period, or two of corn and one of beans. He of course had to adhere strictly to the timetable of sowing and harvesting prescribed by the local agronomist. He also had to be willing to work many more weeks in the year than previously, but his income rose dramatically. After some hesitation, others in the area began to imitate the new technique.

In other areas where none of these innovations are practicable, a major opportunity exists for promoting livestock on mini-farms. After all, until quite recently the majority of the farms in Western Europe were as small as those of Mexico, yet there the operators learned to achieve satisfactory incomes by turning their holdings into mini-factories, using their land almost entirely for feed crops, buying large quantities of supplementary feed and processing it all through the stomachs of dairy cows, hogs, chickens, and rabbits. In Mexico such an evolution has so far not materialized in part because of a traditional disinterest in productive animals. But two other factors have been more important. The first is that ejidatarios' only grazing opportunity is on the collectively owned pastures which the ejido as an organizational unit does nothing to improve. Secondly, the prices of feed grain and other concentrates have been too high relative to the end prices of meat and milk, except under conditions of mass production as with poultry and eggs. Until the authorities come to grips with these two problems, not much can be accomplished in improving labor productivity over large parts of the central regions.

Mention should also be made of a widely held belief that small farmers can be made prosperous by establishing local agro-industries. Obviously the term "agro-industry" can mean either of two things: the establishment of industrial plants in rural areas, a topic to be taken up presently; or, in a narrower sense, the building of food processing plants on the basis of local materials. But nowadays food processing has become a highly technical operation which could not be managed by inexperienced ejidatarios. It requires a steady supply of products for processing, not a few tomatoes one week and a few cucumbers the next, and finally to be profitable it has to be fully mechanized and offers only a small amount of employment. Pioneered largely by foreign corporations, food processing in Mexico is already sophisticated, and there is little likelihood that rustic endeavors could compete against the well-established large enterprises. The pursuit of higher productivity has to be undertaken on the farms themselves.

The conclusions to be drawn from the situation described so far in this chapter are far from optimistic. Except in a few areas of commercial production—the cattle ranches, the larger private farms in irrigation districts and the plantations of the south—overpopulation and underemployment prevail throughout the farm sector, mitigated to some extent by part-time work in other occupations. There exist certain opportunities for increasing the volume of on-farm employment by means of specific attacks on productivity improvement in particular localities, and more generally through a policy of small-unit livestock farming. But simultaneously the steady march of mechanization will continue uninterrupted (even if the government were to desist from encouraging it with subsidies), and this trend will offset, probably more than offset, any employment gains through productivity improvement.

On balance, therefore, the manpower requirements of the farm sector are more likely to decline than to remain stable. Underemployment is likely to persist. Whether its intensity worsens or diminishes will depend, in the long run, on whether and when the growth of population may begin to decline in the rural areas and, in the shorter period of the eighties, on whether the outward migration becomes more rapid—a topic which we must examine separately.

DO IT YOURSELF

Before proceeding to the nonagricultural aspects of employment, another theme deserves consideration. On the one hand, the Mexican villages are overflowing with manpower idle during large parts of the year, while on the other the villages are in a regrettable state of backwardness and the children lack the basic ingredients of a healthy diet. The traveller who observes these things naturally asks himself whether something could not be done to harness the available manpower to satisfy some of the pressing local needs.

The conventional contemporary response to this dilemma is to seek public funds for expansion of public works programs, for support to artisan and handicraft activities or for subsidizing small-scale industries to establish themselves in rural areas. Mexico, like several other countries of Latin America, has a centuries-old tradition of authoritarianism and paternalism. Most of the pre-Hispanic societies were priestocracies or military dictatorships. The Spanish invaders came from a country which had long had authoritarian institutions. After the Independence, and again after the Revolution, the power structure remained the same: merely that a different group of people exercised the power. The peon transformed into ejidatario maintained his habit of soliciting the government for anything desired by himself or his village.

But passivity and paternalism are not enough. Although during the seventies the government devoted large sums to bringing electricity and drinking water to many villages, these are not the only needs, and government has many claims on its limited resources. Also there are various improvements and activities which would not belong in the realm of public works. The choice for the villagers is either waiting until in some indefinite future the government can afford to modernize the villages and/or until with reduced population the village is itself prosperous enough to generate investment resources for its own betterment.

The attitude of government agencies is that for any program a full complement of the latest machinery is required, that it has to be brought to the community from county headquarters and that most of the operators will be headquarters' employees too. It has become almost inconceivable that anything can be done simply. Yet as E. F. Schumacher said: "I should like to remind you that the Taj Mahal was built without cement, electricity and steel and that all the cathedrals of Europe were built without them."[2]

Of course it would be foolish to underestimate the usefulness of

---

[2]Schumacher, E. F., *Small Is Beautiful*, Bland & Briggs, London, 1973, p. 204.

machines. We could not produce the tonnages of corn which the world needs if it all had to be planted with the stick. But neither is it true that we can do nothing until we acquire a bulldozer. To quote Schumacher again: "The output of an idle man is nil, whereas the output of even a poorly equipped one can be a positive contribution." In Mexico's villages, as already emphasized, there are hundreds of thousand of peasants and other rural people who are idle for several months in the year. Simultaneously there is a lengthy list of tasks in villages and on farms waiting to be performed, tasks which would add to the incomes of rural people and to the quality of life in the countryside. The link between available effort and desired satisfactions can be forged only by adopting methods which cut to an absolute minimum the assistance of external resources.

And where should lie the decision-making which would bring unconventional programs into existence? The Chinese with their State-oriented ideology, organized their communes for a multitude of self-help activities, and similarly without much external finance, but all the decision-making came from the center. This was a totalitarian solution. A free society's solution requires the people concerned to decide for themselves what they are capable of undertaking and how to go about it. They may need a little prompting to get started, a little boost to their insufficient self-confidence; yet basically the initiative and responsibility must be theirs.

This approach amounts to saying to the villagers: "Look, there are many things which you want and which you think it impossible ever to get, but you are wrong; ways can be devised for you to obtain at least some of these things." The first step is to induce the people to articulate what they most want in their local situation. The second is to help them distinguish those which are attainable from those which are not. A small group might develop an activity yielding sufficient cash to purchase a bicycle or a sewing machine, whereas the acquisition of a refrigerator or an automobile might have to be postponed until a later stage. Economic realism is something well-rooted in the peasant mentality, even though centuries of paternalism have encouraged him to hope that somehow he might get something for nothing.

The next step is the choice of an appropriate enterprise, preferably one already locally familiar or quickly learned: beekeeping, weaving, hog raising, vegetable cultivation. Where there is an animator his (or her) role should be to steer the choice away from unsuitable projects toward ones likely to succeed. Even the simplest project needs materials, and the villagers have no money. Yet a beehive can be built from odd bits of timber lying around; for vegetable growing an unused plot of land can be borrowed. With income from the sale of honey or vegetables, a few packets of seeds and a few pounds of fertilizer can be purchased, some wool for weaving, and so on.

A hog project is more ambitious. Styes have to be built, though local bricks can be used; water involves a tank and some piping, though neither need be new; hog feed has to be purchased, but credit may be arranged. This implies more organization and, initially, probably more help from outside. Yet there are many Mexican villages in which such projects have been developed successfully.

Some village groups will prefer to give priority to projects which will improve the amenities and services of their village. Most of these are more difficult because they require more materials and do not generate income. Yet a piece of land can be cleared of bushes and roughly levelled into a football field with the aid of a few spades. If there is stone near the village it is feasible to pave the village street with no other help than a wheelbarrow. A swamp can be reclaimed for cultivation by hand digging a few drainage ditches. Some sloping plots of corn land can be terraced and a few yards of pipe inserted in the hillside can bring them water.

Other projects require more expensive purchased materials: a sewage disposal project, enlarging the village school, modernizing local houses with piped water and sanitation. These may be attempted after the less ambitious projects have yielded some surplus income. No first-year violin student tries to play one of the Paganini concertos. It is important to establish in the minds of the participants the direct link between efforts and satisfactions, a linkage often obscured in a technological society where goods apparently come out of machines, or in an egalitarian one in which distributive programs divorce the act of receiving from the act of contributing.

The above-mentioned examples of village projects are not figments of imagination; they are cited from actual achievements in Mexican villages. Quite a number of young Mexicans, university graduates and others, have gone to the villages to spend a few months, or a few years, animating the hitherto inert local people to better their living conditions. A group of such volunteers, each with his or her own specialty, can achieve coverage of such topics as agronomy, animal husbandry, marketing, and sociology.[3] The role of the animator must always be to get the participants to assume responsibility for their projects as soon as possible, partly to stimulate their own pride in their own achievement, and partly to ensure that the project continues when the animator moves away.

Obviously the efforts of a few dozen volunteers in a handful of villages cannot make a serious impact on rural backwardness. The movement would need to be expanded, mobilizing more volunteers, and providing more encouraging conditions of service. Such a transformation would at once court the danger of creating a large bureaucratic organization which would pass into the hands of some government agency and reestablish the paternalism of the past. For rural do-it-yourself activities to succeed in a country like Mexico, they must preserve their spontaneity and voluntary character, with no more organization than, for example, that of the Red Cross.

Nonetheless, given the pyramidical hierarchy of Mexican society, it would be necessary to obtain a blessing from the highest level, the president himself; otherwise the movement would encounter lethal obstructionism right down the line to the level of the political bosses in the villages. Their approval, at least tacit, would be a sine qua non.

---

[3] See in this context: *The Role of Social Science Research in Rural Development: a View from Mexico*. S. I. Friedmann, Rockefeller Foundation Conference, New York, 1975.

Rural self-help appeals to people of many shades of opinion. For those who believe that society should set about creating Socialist Man, it offers opportunities for awakening a social sense through the practice of cooperation. For those who believe in the virtues of individual effort it offers scope for mobilizing productive initiative. For those who believe that democracy has to be learned at the grass roots, it offers the reality of decision-making by consensus.

It would be foolish to exaggerate the potential of these proposals. Even a generous multiplication of what has been tried out in a few villages will not transform the structure of Mexico's rural society. Yet because the more conventional approaches to rural poverty remain so inadequate, this alternative is worth a trial. Human beings are the most precious resource society possesses, and their neglect and underutilization constitutes a confession of failure. We cannot harness human capacity more fully by creating a new government agency nor by writing yet another chapter in the Five Year Plan. And we do not need to wait until vast financial resources are available to dedicate to grandiose programs. A start could be made here and now to provide productive work for several thousands of families, thereby achieving worthwhile additions to individual and social welfare.

NONAGRICULTURAL EMPLOYMENT

From these practical, but inevitably somewhat limited, proposals we must now turn to consideration of some of the more macroeconomic aspects of rural underemployment, since most of the solution to this problem in the medium and longer term lies outside the farm sector. It is true that in a sense the farm population can itself make a contribution by gradually diminishing its own rate of growth; but family planning reaches rural areas last of all, and in Mexico one cannot count on much from this factor during the immediate future. So we are left with the question: how much nonagricultural employment can be generated for the benefit of farm people?

Yet before proceeding to answer this question directly, we must give attention to the rather special nature of peasant demand for employment. Some people imagine that the provision of more jobs for farm people amounts merely to offering work in a factory or store or office with the normal duration of 40 to 48 hours five or six days per week. This is much too crude. Peasants have a radically different tradition of work: from time immemorial they have adjusted their rhythm to the seasons, working very hard indeed when, as at harvest, much has to be accomplished in a short space of time, and accepting idleness as something normal during periods when no work is required in the fields. For them work is, by its nature, something intermittent.

Many small-scale farmers and their sons, in Mexico as elsewhere, undertake off-farm work in their spare time; but their first responsibility is to their land and its needs. Hence, unless they are migrating out of agriculture altogether, any other work which they accept must be of a kind which combines with their agricultural commitments. This in practice means part-time work at not too great a distance from their homes.

Unfortunately, there exists only fragmentary evidence as to how far this phenomenon has developed. A part, but only a certain part, of the answer can be obtained from the data in the occupational section of the population census which ascertains the "principal occupation" of the informants. Thus, part-time farmers may tend to class themselves as bricklayers, truck drivers, etc., if they work five days a week in these occupations returning to their farms at weekends and/or taking time off from their nonfarm work to attend to the sowing of corn and to the harvest.

Comparing the latest population census of 1970 with that of 1950, we find that while during those years the farm population as a whole increased by 5.8 percent, there were twelve states in which numbers actually declined. The majority of these were states containing one or more large cities offering employment opportunities to peasants living within commuting distance, while the other states were ones of marginal farming where extreme poverty forced families off the land—factors exemplifying the pull and push effects of economic circumstances.

Another source of evidence comes from a special survey undertaken by the Center for Agrarian Investigation in contrasted rural areas. In a poor district in Oaxaca remote from cities the ejidatarios worked on their parcelas only 48 percent of a normal working year and did little else in their idle time, though some of their family members did obtain remuneration off the farm. Quite different were the findings in the Valle de Toluca (state of México) within commuting distance of both Toluca and Mexico City, where the operators devoted only 10 percent of their time to their land and where they and their family members obtained more than half their income from nonfarm activites.

A different class of employment is that obtained by migrants, legal and illegal, to the United States. The great majority of these are from farm families suffering from underemployment and most of them stay away only part of the year, arranging to be back home in October-November to help with the harvest. From some villages, even though very poor, no one ever makes the trek, while from others in the same district annual visits to the U.S.A. have become a tradition. In these latter the physical consequences of the repatriated savings are everywhere visible: houses rebuilt and fitted with modern conveniences, trucks brought back and used for local transportation services, gas guzzling automobiles albeit of ancient vintage.

For obvious reasons no reliable data exist on the volume of work obtained by the braceros, but as regards nonfarm employment in Mexico we do have data from the censuses. Using the concept of economically active population, that is those normally in employment though at the date of the census they may be sick or temporarily unemployed, the nonfarm sector's employment increased by 127.7 percent between 1950 and 1970 while the population as a whole was increasing by 87.6 percent, no mean achievement.

However, off-farm employment was expanding at very different speeds in different parts of the country. For instance it sextupled in the state of México as a result of an overflow of Mexico City's industry into that state, and it quadrupled in northern Baja California through the growth of the frontier towns. At the other extreme it increased by only 34.7 percent in Yucatán and only 47.5 percent in Oaxaca. It is believed that these trends continued during the seventies, with the difference that other cities, in addition to Mexico City and those of the frontier, were becoming poles of attraction. As a consequence the farm population as a whole has probably ceased to expand and is indeed declining in the better communicated regions of the country.

THE MEDIUM TERM EMPLOYMENT OUTLOOK

It is not enough for the farm population to stabilize, because that will still leave a substantial volume of underemployment, and also, as we have seen, the agricultural demand for manpower is likely to be declining during the eighties. For rural incomes to rise, more employment is needed for farm people, preferably located in rural areas and preferably to a large extent part-time in character. To evaluate the chances of this happening, we must evaluate in broad terms the prospects for employment as a whole.

In such an exercise it is customary to start from the number of persons of working age, usually described as the "labor force" and defined as all persons between the ages of 15 and 65 (though in a developing country many persons above and below these limits may be at work). Within this age group a considerable proportion is not available for employment, i.e., the full-time housewives, young persons still in full-time education, persons permanently incapacitated, etc. The remainder constitutes the "economically active population" (EAP). The ratio of this EAP to the labor force, known as the participation rate, varies from country to country; in Switzerland and Japan it is as high as 70 percent, due largely to high female participation; in developing countries it is generally much lower, in Chile only 46 percent. In Mexico in 1970 the participation ratio was 51.8 percent, one of the lowest in Latin America, partly explained by the tradition, still strong, that females should not work outside the home.*

With the expected decline in the population growth rate during the eighties, the number of persons of working age as a percentage of the total should increase slightly. Also a slight increase in the participation rate should be expected, reflecting an increase in female employment as shown in Table 10.2. If this is projected to rise to 53.8 percent then the EAP of 1990 would total 24.5 million, compared with 17.8 million in 1980.

These estimates and projections imply that during the seventies job creation proceeded at an average annual rate of 3.0 percent, a considerable improvement on the 2.5 percent achieved during the sixties. Nevertheless, the participation rate did not improve at all because this was the decade of maximum growth of the labor force, and large numbers of adult persons willing to work were unable to find any. Moreover, it is doubtful, in view of

---

*Revised figures are used to accord with the official revision of total population.

TABLE 10.2

ECONOMICALLY ACTIVE POPULATION, 1960 TO 1980;
AND PROJECTIONS TO 1990

	Popu-lation	Labor Force	Economically Active Population	Partici-pation Rate	Popu-lation	Labor Force	Economically Active Population
	Millions			Per-centage	Annual Growth Rate Percentage *		
1960	36.0	18.2	10.3	56.5	3.5	3.4	2.5
1970	50.7	25.5	13.2	51.8	2.9	3.1	3.0
1980	68.0	34.5	17.8	51.6	2.7	2.8	3.2
Projections:							
1990	88.4	45.5	24.5	53.8			

SOURCE: Statistical Office, population censuses for 1960 and 1970 data. Author's estimates for 1980 and 1990.

* Growth rates are: 1960–1970, 1970–1980, and (projected) 1980–1990.

the economic crisis of 1975–76, and the slow recuperation thereafter, whether the estimated 1980 figures were really attained.

As to the eighties, the Mexican government has been placing more emphasis than ever before on employment expansion, setting a target growth rate of 4 percent per year. The general impetus to the economy provided by petroleum production should also help in this direction. Assuming that reality falls a little short of the employment target, a growth rate of 3.2 percent during the decade would still permit the small improvement in the participation rate to the 53.8 percent just postulated.

In practice there is, of course, a trade-off between growth in employ-.ment and growth in labor productivity, at any given rate of growth of domestic product. For instance, during the sixties gross domestic product increased on average at the high rate of 7.1 percent, but employment expanded relatively slowly at 2.5 percent permitting productivity to grow at 4.6 percent overall. During the seventies the situation was reversed. With GDP growing rather more slowly at 5.7 percent, employment increased quite rapidly (3.0 percent), and consequently labor productivity improved only at 2.7 percent per year.

When making demand projections in Chapter 2, we offered two alternative GDP growth rates for the eighties: a high of eight percent and a low of six percent. For our present discussion of employment, we may fix on a single projection, and so postulate a seven percent GDP growth rate. With this assumption and assuming also that the employment growth of 3.2 percent is achieved, the growth in labor productivity would be 3.8 percent annually. This would be considerably below the rate attained in the sixties, but much better than the performance of the seventies when apparently productivity was sacrificed to employment objectives (Table 10.3).

More light can be thrown on the plausibility of these projections by examining the past trends of employment and productivity in each of the three

TABLE 10.3

**ECONOMICALLY ACTIVE POPULATION BY SECTOR
AND PRODUCTIVITY, 1960 TO 1980; AND PROJECTIONS TO 1990**

(annual growth rates)

Period	Agriculture	Industry	Services	Total
	Gross Domestic Product			
1960–70	3.7	8.8	6.8	7.1
1970–80	1.5	6.9	5.6	5.7
1980–90 (projections)	3.0	9.0	6.0	7.0
	Economically Active Population			
1960–70	0.6	3.0	4.6	2.5
1970–80	0.3	3.6	4.5	3.0
1980–90 (projections)	−0.5	4.5	5.0	3.2
	Labor Productivity			
1960–70	3.1	5.8	2.2	4.6
1970–80	1.2	3.3	1.1	2.7
1980–90 (projections)	3.5	4.5	1.0	3.8

SOURCES: Gross domestic product, growth rates 1960 to 1980 from Banco de México, *Informe anual*. "Economically active population" growth rates from Table 10.2. Labor productivity equals the difference between the two previous items. Projections from 1980 to 1990 by author.

main sectors of the economy. During the sixties, a period of vigorous industrialization in Mexico, employment grew very little in the primary sector (which principally means agriculture, since petroleum is classed as part of industry), fairly well in manufacturing and rapidly in the service sector. As to productivity, this rose quite satisfactorily in farming because the growth in production did not begin to level off till the later years of the decade. In industry, the new plants being erected were modern and mechanized, facilitating a prodigious rise in productivity at the expense of more employment. In services the productivity increase was only moderate.

During the seventies the employment picture closely resembled that of the previous decade, except in manufacturing where the growth rate improved from 3.0 to 3.6 percent. But because of changes in sectorial GDP growth rates, the productivity picture was different. Its growth rate declined steeply in all three sectors, which is consistent with other evidence showing that real incomes were rising much more slowly than before. Whether productivity actually declined in the services sector is open to doubt: if when the definitive figures become available, it emerges that employment expanded at less than 3.0 percent overall, then the shortfall most likely occurred in the services sector (where anyway the statistics are weakest).

Against the background of these experiences, the projections by sector in the eighties have been formulated. In agriculture the volume of employment may be beginning to decline. Its GDP growth we have projected at three percent, which corresponds with the 3.5 percent growth in *gross* output postulated in Chapter 6, and is optimistic. The resultant productivity growth rate at 3.5 percent is higher than in the sixties.

In industry, including petroleum, we have set a high employment target of 4.5 percent annual increase, because otherwise it would be impossible to achieve the overall target growth of 3.2 percent without overmanning the services sector even more than it is at present. This target implies a modest recovery in the productivity of industrial workers, yet an increase still far below that of the sixties. This situation is probably because, during the eighties, a large proportion of the industrial expansion will consist of oil refining, petrochemicals, electricity, machinery and other capital goods—all of them capital- rather than labor-intensive. Indeed, if the overall employment target of 24.5 million jobs in 1990 fails to be met, the shortfall will certainly occur in the manufacturing sector; or, alternatively, it could still be met if more manpower flows into the services sector than we have postulated.

On the whole, the scenario described above may err on the side of optimism in several respects. For instance, if inflation remains high, it may prove impossible to reconcile this with a GDP growth rate as high as seven percent. If the public sector continues to expand at the expense of the private sector, this might be a trend favorable to employment but unfavorable to productivity, because of the overmanning which characterizes public enterprises. What may well occur is a widening of the productivity differences as between different branches of industry: for example, rapidly rising productivity and incomes in the petroleum, electricity and petrochemical groups and only slow increase elsewhere. Such changes might give rise to social tensions and labor unrest. It follows that the employment outlook cannot be regarded complacently, in spite of the expected petroleum boom, and that policies other than those of general economic expansion need to be formulated. Since these are relevant to generating more jobs for redundant farm people, a few examples of what might be attempted are given in the following section.

IMPROVING THE MANPOWER DEMAND

In most developing countries one of the principal constraints on rapid economic expansion is the shortage of investment resources. The transformation of these countries into industrial societies has come at a moment unfortunate in the sense that modern technology, much more than nineteenth-century technology, requires a large investment of capital for every extra unit of output and extra unit of employment. This means that industrialization requires a lot of what the third world lacks, namely capital, and uses little of what it has in abundance—manpower.

It is true that since 1973 some of the petroleum producing countries have contrived to accumulate enormous surpluses of capital, so that in the Arab countries at least it is skilled manpower which has become the bottleneck. Others, however, for example Venezuela and Indonesia, have also generated capital, but it proved insufficient, or was inadequately used, to promote industrial development. In Mexico the petroleum of the eighties will certainly generate financial resources in large volume, but there are several reasons why they may be inadequate. For one thing Mexico has a large population, much larger than that of any other oil producer except Indonesia. Secondly, she will need to spend heavily on food imports, and will have a strong propensity to import all manner of other goods, especially if the authorities are reluctant to make adjustments in the rate of exchange. Thirdly, the branches of industry scheduled for expansion are mostly capital-intensive. Fourthly, most of them will belong to the public sector, notoriously wasteful in its resource use. For all these reasons, Mexico is likely to continue to feel capital constraints, and hence it will be necessary to adopt policies which seek to increase the employment coefficient for any given input of investment.

One of the ways in which investment resources are wasted is in the underutilization of productive equipment. Though in this respect Mexico's performance is much better than the third world average, it is far from satisfactory. An inquiry late in 1979 revealed that manufacturing industry as a whole was working an average of 1.7 shifts per day; but some branches were working only 1.2 shifts, for instance pharmaceuticals and furniture making.[4] To improve this situation efforts are needed in several directions simultaneously. Amplified training programs could be initiated to reduce the bottlenecks caused by shortages of skilled workers. Government could simplify the delay-provoking formalities in granting licences for the importation of spare parts and other supplementary equipment. Tax incentives could be granted, e.g., more generous depreciation allowances on equipment used for two or three shifts. The numerous national holidays could be transferred to Sundays to reduce time lost.

Another broad field is that of choice of production technique. While in a few types of industry very little choice exists, in the majority of cases the management can opt for more capital- or more labor-intensive methods, the decision resting on external circumstances which government can influence. One of these is the persistent distortion between the factor prices of labor and of capital. In Mexico where half the labor force is unionized, the workers obtain rates of pay which may not be high in terms of purchasing power but are often exorbitant in terms of productivity levels.

Moreover, apart from union pressures, Mexico has a comprehensive system of statutory minimum wages covering almost all manual occupations of significance. In many instances these minima are set at levels well above the productivity of many persons who nominally belong to those occupations, with the consequence that either they are not offered employment or the employer evades the law and pays them less.

Another deterrent to increasing the volume of employment lies in the manner in which firms are rated for social security contributions, of which

---

[4]Banco Nacional de México, S.A., *Examen de la Situación Económica de México,* October 1979, p. 349.

there are many. These are levied in proportion to the payroll, i.e., the more workers the firm has the more it pays—evidently a strong inducement to increase mechanization wherever feasible.

Furthermore, mechanical equipment is relatively cheap. Much of it is imported from abroad with licenses readily granted and free of duty, and from countries whose inflation rate is lower than that of Mexico. It can be paid for with money borrowed at negative real rates of interest. It does not answer back, does not go on strike. There is a Mexican saying: "The fewer the workers, the fewer the problems."

To counteract these antiemployment influences, consideration might be given to levying import duties on automated equipment not strictly necessary, to raising interest rates to levels where they become positive (i.e., above the rate of inflation), to increasing the export incentives for classes of goods produced by labor-intensive methods, and so on.

Another positive move would be to reinforce the steps already taken to encourage the decentralization of industry to provincial locations. This would have two advantages. It would stimulate the volume of employment, because industry in the provinces tends to be more labor-intensive. Also it would generate employment in locations physically accessible to the underemployed farm population. Already the Mexican government has nominated certain towns as its selected poles for future industrial development, and has offered subsidies and tax incentives to firms establishing plants there while at the same time making more difficult the expansion of plants in the overcrowded metropolitan area. However, the sticks and carrots are still not large enough. Moreover, firms will continue to find provincial locations unacceptable so long as they have to come to Mexico City to obtain permits from one or other government agency for their every activity. Only a radical reduction in the complexity of the permit system, or a decentralization of the permit-granting agencies themselves, will really bring decentralization alive.

But, as our earlier analysis revealed, it is not enough to bring industry and employment to the smaller towns. The form of the employment, or at least of some of it, has to be adapted to the rural families' need for part-time work—either part of the day or part of the year. In some countries certain firms have been able to arrange the making of components in ways that allow the work to be distributed among peasant families working in their homes. Other firms deliberately organize handicraft activities of which they supervise the design and quality standards, and subsequently effect the marketing of the output. In Mexico much remains to be tried in both respects. From Japan and South Korea techniques could be learned of utilizing village manpower in electronic and microprocessor components. In artisan products, which have hitherto been oriented almost entirely toward the North American tourist, there remain untapped national markets for a wide range of village products provided that more attention be given to adapting traditions to the requirements of today's consumers.

Apart from the various opportunities for adapting the employment characteristics of manufacturing industry to suit rural needs, adaptation is also required in the organization of public works programs. In Mexico the provision of rural electricity services, water supplies, village housing and so on, tends to be organized from at least the state capital and at times direct from Mexico City; so that one may see truckloads of workers being transported fifty or sixty miles to work in a village while the local inhabitants sit around as idle spectators. Admittedly it takes more ingenuity to mobilize local people to participate in these programs, but in the end it saves money as well as giving them activity. And shortage of investment funds will remain the limiting factor in public works projects. What has been accomplished in recruiting local people to build rural roads could be applied to the provision of other services.

IMPACT ON THE FARM POPULATION

The preceding sections of this chapter have been concerned with a review of the overall and sectorial employment prospects for the future and with how these could be made even brighter through official action. The prospect in general is more hopeful than in any previous decade, yet still the year 1990 will see a low participation rate and much continuing underemployment. There are few signs that government will try to restrain the trend toward more and more capital-intensive production techniques. Legislation has been enacted, it is true, to push industry toward provincial locations, but it is too early to judge how effective it will prove in practice.

We need to recall that the rural job hunters fall into two groups: those who are prepared to quit their farms and their villages and take their families to the cities in search of work; and those who desire supplementary work, but only on a part-time basis (which means at home or in an accessible location nearby). Employment policy for rural areas has to provide for both groups.

As to the first, the decentralization of industrial growth will hopefully orient their migration away from the metropolitan area, but even in these other cities they may not find jobs when they arrive because they lack the qualifications for the employment available. What is needed is a massive expansion of vocational training directed especially toward newly arriving migrants. Otherwise, they will drift into the already overmanned occupations of petty trading or add to the ranks of the urban unemployed.

The second group is undoubtedly more difficult to help. The government hopes to make a contribution by its programs of special aid to small and medium-sized enterprises—low interest credits, tax holidays, etc. But enterprises do not, just because they are small, provide employment on the basis of two to three days per week or six to eight months in the year. It should be feasible to create incentives especially for firms willing to organize the pattern of their employment on such a basis. The result would be a bettering of rural incomes comparable to that which is attempted by the frequent increases in farm product prices, and without the disagreeable inflationary effects on the cost of living. In certain quarters there might be opposition to such a policy, denigrating it as an encouragement of sweated labor among village people, and some unions might see therein a threat to their privileged position. But it would be in the general interest that such objections be countered and, if necessary, overridden. Not every un- or underemployed person can be given employment in the next years, and meanwhile low productivity and low pay is preferable to zero productivity and no pay.

We also have to look further ahead. The volume of farm employment has been projected as almost stationary during the eighties, for the first time after decades of continuous rise. This will be the watershed before in the nineties begins the inevitable and equally continuous decline in farm man-power requirements, which characterizes every country that moves further along the road to prosperity. It is quite likely that this decline in requirements will set in earlier and more rapidly than any natural decline in the farm population brought about through the spread of family planning, in which case the pressure on farm families will become more severe to seek nonfarm employment either locally or in the cities or in other countries.

Hence arises the urgency of employment policies tailored to the needs of these people. It will be difficult enough in Mexico to meet the mac-roeconomic employment targets in a decade during which expansion will be based on petroleum and the high technology industrial activities arising there-from. Considerations of prestige will be impelling the authorities to opt for nothing but the best—which in practice is thought to mean copying the most sophisticated techniques of Big Brother. It will be much harder to contrive sufficient encouragement for the types of industry in which farm people without farm work can participate—the skills necessary, the feasible loca-tions, the combination of a new occupation with residual land cultivation.

Yet without a determined effort to succeed on this front, the talk of income redistribution and social justice will continue to sound hollow and unconvincing. It is of course possible that a petroleum-rich country could indulge in the luxury of an ambitious expansion of welfare services, which would guarantee modern living standards to every citizen, rural as well as urban, without any obligation on his part to engage in productive work. Nonetheless it is doubtful whether, with a population in 1990 on the verge of 90 million, Mexico would have wealth enough to pursue this course. Moreover, it is certain that such a policy would not create the kind of society to which the Mexican people aspire.

If this conclusion has validity, then employment policies will have to be rethought and reformulated. They need to be designed, much more specifi-cally than hitherto, to meet the needs of the millions who are going to have to change their occupations, and in many cases their habitats. The task is not an impossible one, but it requires a more imaginative understanding of the men-tality of the un- and underemployed, particularly of the farm people. Their incorporation into the new and prosperous Mexico can be brought about, not through welfare, but through facilitating their participation in many different types of productive employment; not through accepting as inevitable their mass migration to the two or three gigantic cities of the Republic, but through insisting that economic activity be sited in provincial poles of growth—first a few and later several hundred. It will take time, and only a beginning can be made in this present decade but, once such policies are adopted and im-plemented, their benefits will be apparent for many decades to come.

11. Farming in a Petroleum Economy

Mexico faces an agricultural dilemma. On the one hand, the locomotive impulse to production expansion has run out of steam; on the other the demand for farm products is rising faster than at any previous time. For several years now this imbalance has preoccupied political leaders and economic analysts, for it is not a dilemma that will quietly go away of its own accord. It has already persisted since the mid-sixties and will continue unabated unless something is done to resolve it.

There are, of course, several ways in which the problem could be dealt with, and to a major extent the choice which Mexicans make will depend on how the country's economic and political climate evolves during the eighties. That is why in this final chapter the analysis will be stretched beyond the confines of the farm sector to consider some of the factors which will be determining that climate and hence the likely characteristics of agricultural policy.

AN OIL BONANZA?

Most people, if asked what feature may be expected to dominate the Mexican scene in the immediate future, would reply with one word: oil. Much publicity has already been given to the volume of proven and potential petroleum reserves, yet every few months new and higher figures are announced. Of course in Mexico, as in any oil producing country, it is easier to make conjectures than to obtain hard facts; nevertheless, without a commitment to specific figures, it is generally agreed that the country possesses a great deal of oil and that most probably the reserves are higher than anyone yet suspects.

Since the late seventies production has been increasing rapidly and by 1980 had already surpassed two and a half million barrels per day. Exports exceeded one million barrels per day and were rising rapidly. Some people expect these figures to be doubled long before the end of the decade. At the same time a new petroleum refining capacity is being installed, so that at least part of the exports can take the form of refined products. New petrochemical plants are being built and old ones extended. Expansion is under way for other downstream products such as artificial fibers and fertilizers.

Pemex, the State petroleum corporation, has become a major purchaser of a wide range of equipment. Therefore, both government enterprises and private firms have a strong incentive to develop production lines which will meet at least part of that demand, and they in turn need more machinery and other equipment. This has given a fillip to most of the capital goods industries, which in turn has had favorable repercussions right down to the raw materials required. Thus oil is already stimulating new activities in many sections of manufacturing industry, a stimulus expected to gather momentum during the decade.

On the financial side, the ever-rising world oil prices coupled with the increasing volume of Mexico's oil exports are generating each year larger and larger foreign exchange earnings. Furthermore, since the government takes for itself 52 percent of these earnings in the form of an export tax, as well as other taxes levied directly on Pemex itself, the oil expansion massively augments government revenues. In official thinking, it is expected that this will make a major contribution to financing the very considerable budget deficit which during the seventies became a permanent feature of public finances. In sum, the oil boom is counted upon to achieve favorable developments in manufacturing industry, in foreign trade, and in the fiscal situation.

However, there is another side to the coin. First, and very important, Mexico, unlike all other major third world oil exporters except Indonesia, is a country with a large population—already nearly seventy million in 1980. This implies that what looks like a large oil industry and voluminous benefits, turns out to be much more modest when reckoned in per capita terms. This may be illustrated by the following comparison. If Saudi Arabia, the world's largest supplier, were exporting nine million barrels per day, Mexico in order to attain the same export value per capita would have to export 70 million b/d compared with the one million she was exporting in 1980.

Because Mexico has a large, and still rapidly increasing, population, and is at a comparatively early stage of development, she has many pressing needs for public investment in infrastructure, for example ports and transportation facilities, as well as for social services of all kinds. For years to come Mexican governments will always have more demands for investment than means to satisfy them; hence the public sector deficit may not disappear.

Because the population is large, because per capita incomes are rising, and because industry depends on external sources for a considerable portion of its raw materials and equipment, the propensity to import is very high. In the past, though imports considerably exceeded exports, the gap could not become excessive since it was limited by the availability of loans from abroad. The consequence of rapidly rising revenues from petroleum exports may be not to eliminate the deficit on current external account, but rather to facilitate and finance a large expansion in imports. Indeed, government thinking presupposes that foreign borrowing will continue to be necessary throughout the eighties to finance both the continuing import surplus and part of the domestic investment needs.

A rather different reservation about the oil outlook relates to the likelihood that the Mexican government may decide to move slowly in expanding its drilling operations. It has read its lessons from what happened in other countries which encouraged a skyrocketing of oil production, failed to employ the profits wisely, and subsequently faced declining output. Not only does Mexico want its reserves to last a long time, but also it wants to exploit them at a rhythm which will permit the proceeds to be absorbed into the rest of the system productively and without too much overheating. It wants to avoid becoming a one-product country and instead use this oil resource to diversify generally and strengthen its economy.

In any case, and even despite the best intentions, the non-oil sectors have begun to experience difficulties, and these can be expected to become more acute. Pemex and its principally supplying industries have already made heavy demands on available investment resources, causing a restriction on the volume of credit available to the rest of the economy. Pemex and its suppliers need a large number of skilled workers whom they attract with relatively high wages, creating a shortage of skills in other sectors. Rising wages, high interest rates, and the continuing inflation are discouraging to certain branches of industry, particularly those which, because of price controls, cannot pass their cost increases on to their consumers. And when inflation is coupled with government reluctance to adjust the value of the peso, the firms which rely to any appreciable extent on exports find themselves progressively disadvantaged.

In more general terms the private sector sees its role diminishing while that of the public sector expands. There are several reasons for this, as we shall see, but the oil boom is one of the more important. Pemex itself and most of the basic petrochemical plants are State enterprises projected for rapid expansion; also the substantial tax revenues deriving from Pemex operations will enable the government to finance more and more public sector enterprises both in manufacturing and in infrastructure. Already before 1980 the annual volume of new fixed capital formation was larger in the public than in the private sector, and this public sector predominance is destined to increase. Furthermore, during a period of continuing inflation, governments try to mitigate its adverse effects by tightening up their price controls and by other devices, all of which squeeze the profits of private sector enterprises, which in turn reduces the volume of their reinvestment and finally diminishes their capability of maintaining their role in the mixed economy.

More specifically, in regard to the farm sector, the first consequence of petroleum expansion has been and will be to increase the volume of purchasing power in the hands of the general public, even though in real terms this is attenuated by inflation. With more money in their pockets, consumers buy more household durables, automobiles, etc. but they also buy more food and particularly more of the expensive foods. On the other hand, the oil boom does nothing directly to promote an expansion of agricultural production, which might help satisfy the rising demand. It is possible that the government with more financial resources at its disposal may come to devote larger sums to major projects in irrigation, land reclamation etc., but the private agricultural investor will find loan funds scarcer and more expensive than hitherto and will experience no reduction in the many disincentives which have been identified in previous chapters.

All in all, the fact that oil will dominate the Mexican economic scene during the eighties must be expected to produce a mixed bag of consequences for the rest of the economy. There will be stimuli but obstacles, rising demand but rising costs, shortages and surpluses. Petroleum is not going to provide, as some fondly hope, an easy and instant solution to all the nation's problems. Whether the negative or the positive influences predominate will depend on other factors in the political and economic scenario to which we must now turn.

GRANDPA MARX

As is well known, Mexico has enjoyed remarkable political stability for more than fifty years, a record unmatched by any other Latin American country. The only serious aberration from this pax politica occurred in the second half of Echeverría's presidential term leading up to the crisis of confidence in 1976 and the devaluation of the peso. President López Portillo set about restoring confidence and rehabilitating the economy, a task at which he succeeded in the short space of two years. He pursued middle-of-the-road policies, reverting to that watchword of earlier presidents: conciliation. He prevented political pressure from festering into political confrontation; he sought consensus and obtained cooperation.

But whatever President López Portillo's virtues, he will not be the only shaper of the political scene in the eighties: new presidents will be elected in 1982 and in 1988, and what their personalities will be is anyone's guess. It is true that the PRI, the governing party for five decades, is supposed to act as the integrator of the many currents of opinion emanating from different sections of the community,but in real life it is more often the president who guides the PRI than the other way round; his personal predilections can be decisive, even though he himself has to be sensitive to pressures.

In 1979 President López Portillo put through the Chamber legislation known as the Political Reform which for the first time allowed a considerable number of political parties to obtain registration and participate in the election of that year, including the hitherto banned Communist Party. As a result, several parties of the left secured representation, although in practice a very small number of seats when compared with the continuing overwhelming predominance of the PRI. The most significant feature of the innovation was that no serious party of the right made its appearance. The middle and upper-middle classes and the business leaders continued as before to avoid political participation.

This omission, superficially justifiable in the short-term because they count on the PRI to look after their interests, may have nefarious consequences in the longer run. If the PRI were flanked by parties both on the right and on the left, it could maintain its middle-of-the-road position as general conciliator. If, as at present, its only opposition comes from the left, then conciliation and consensus imply a gradual shift of its intellectual position toward something left of center. And indeed, actuated by the same motive, the president incumbent will be likely himself to adopt such a modification of alignment and use the PRI to put it across. Thus, if the presently existing pattern of parties persists, the eighties may see a gradual shift toward more government intervention, more public sector dominance, in a word, more Statism.

There is also another factor operating in the same direction, one which stems from the coloration of the institutions of higher education. For more than thirty years in all the major universities the faculties of political science, law, economics, sociology, history, and literature have been dominated by socialist thinking and the teaching body has been almost entirely Marxist. Nor is there any present sign of weakening in this ideological monopoly. As a consequence, the senior cadres have a socialist outlook, except for some few who have studied abroad and returned with other persuasions. These people dominate the media—press and television—and also the upper echelons of the civil service where they tender their slanted advice to ministers and the president.

Another group of persons is concerned more altruistically with the large inequalities of wealth and income which exist in a developing society such as that of Mexico. To them it appears that the present system makes the rich richer and does little to relieve the poverty of the poor. For them, the implementation of social justice means placing emphasis on (egalitarian) distribution rather than on production; indeed they are prone to advocate distributing more than has been produced. For them any social or economic problem that arises can best be resolved by giving more responsibilities to the State. Not unnaturally therefore these persons join forces in practice with the Marxists.

It is sometimes said in Mexico that among the political leadership 90 percent of the talk is socialist but 90 percent of the action capitalist, and that the private sector continued in vigorous health. This is up to a point true, but it is also true that whenever institutional changes are made they are in the direction of greater State intervention, more public sector activity, and never the reverse. In Mexico the public sector is not yet as large as it is in several so-called advanced democratic countries, but during the eighties its size will increase, and this trend will meet with the overt or tacit approval of the majority of the educated population.

PATERNALISM VERSUS ENTERPRISE

This inclination derives partly from traditions rooted deep in Mexican history. The pre-Hispanic governments were religious and/or military dictatorships: the Spaniards brought with them their own concepts of. authoritarianism. Moreover, because they encountered a large indigenous population they could settle down to giving orders without having to work themselves, whereas the Anglo-Saxon settlers further north had to do everything with their own hands and created their ethic of working for reward.

In Mexico, the belief and experience of many is that rewards come less from working and more from lobbying or alternatively from asking a favor of the Virgin of Guadalupe. It was noted in earlier chapters how ingrained is paternalism in the ejido section of the farming community. The ejidatario defers to the authority of the ejido's president, who in turn ingratiates himself with the political bosses at county level to obtain favors and personal advancement; and so on up the pyramid of power. An ex-secretary of agriculture said that the ejido was an institution ill-adapted to production but excellent for getting out the votes.

On another plane the elaboration of bureaucracy in modern society discourages individual initiative and strengthens paternalism. It weakens pluralism and abets the concentration of power. In Mexico during the past decade the size of the bureaucratic machine has grown rapidly as its responsibilities have multiplied. In agriculture President Echeverría transformed the directorate of Agrarian Affairs into a separate federal department, and thus it remains although its tasks are coming to an end. In the Department of Commerce there has been an ever increasing proliferation of the number of committees dealing with price fixing and price controls. It is significant that President López Portillo, when faced with an overheated economy whose main symptom was a large budget deficit, could not bring himself to cut back on the bureaucracy; on the contrary, the costs of the administration continued to expand in real terms.

Certainly, in Mexico, as in other third world countries, private enterprise has a bad reputation. Entrepreneurs for the most part prefer to sell a little for a lot rather than a lot for a little. Toward their workers they tend to behave like the rugged capitalists of a century ago. Too many of them use their profits for ostentatious living rather than for investing in business expansion. The profit motive and the profit system are identified with Uncle Sam. Hence a good anti-American is also an anticapitalist. The ugly face of capitalism is identified with transnational corporations, which most Mexicans fondly believe are exclusively American. In agriculture the phobia at the time of the Revolution was directed against the giant landowners, and it is still directed against the larger ranchers and crop farmers, even though these produce most of the nation's food. This is a principal reason why it is so hard to obtain a hearing for proposals which would provide security of tenure for this class of operators.

Individual enterprise and initiative are suspect partly because certain individuals, too many indeed, have behaved badly; partly also because when they compete directly with public enterprise, as for instance in banking, they tend to be more successful. Significantly, the only proposals seriously entertained for reforming the ejidos are ones such as collectivization in one form or another, which would reduce not increase the scope for initiative on the part of individual ejidatarios. Hardly anyone advocates permitting the best of them to become independent farmers.

In the coming years life will continue to be made difficult for individual entrepreneurs and for the private sector in general. The system of permits which are required in so many business activities will be maintained, and may be extended if, in a persistently overheated economy, scarcities develop of certain materials and of services such as freight transportation. Payroll taxes will probably be increased to help finance more ambitious social welfare programs, although by their nature they discourage the expansion of employment. Credit restrictions will remain in force and the banks' compulsory deposit percentage, which the Bank of Mexico uses for financing public sector enterprises, will likely remain above 40 percent. Continuing inflation and low rates of savings will keep interest rates at nominally high levels. To prevent the rising tide of production costs being passed on in full to consumers, the government will multiply the number of articles subject to price control and thereby squeeze profits. This policy, clumsily administered, was already causing shortages of certain foodstuffs in 1980; instead of being protected the consumer was being penalized. Situations of this kind will probably multiply.

The alternative would be for the government to summon up courage to stage a dash for freedom. Although it would be impractible to abolish all price fixing at a single stroke, the market could be set free step by step over a period of, say, two years. Admittedly, the initial consequence would be a general rise in the prices of the goods liberated as profit margins were restored to acceptable levels. But it is legitimate to expect that after a time production becomes again attractive, its volume increases, production costs decline, and the benefits pass down to the consumer who ultimately enjoys a wider choice at lower prices than under the previous regime of controls. It would be unrealistic however to expect developments of this kind in Mexico in the eighties.

THE SPATIAL ILLUSION

Mexicans share with a number of other peoples the illusion that any difficult problem can be solved by multiplying the number of government agencies and giving them more power. But Mexicans also have a special illusion of their own which relates directly to agriculture and that is the illusion of unlimited space, that their country has land enough for everybody and for every desired purpose. They drive the 3,200 miles from Tijuana to Puerto Juárez, passing through few towns and only three or four cities, beholding to right and left vistas of empty countryside and are overwhelmed by the vast resource potential that apparently remains to be used. How can it be other than lack of organization which causes the nation to be importing millions of tons of grains, oilseeds, milk powder, and other foods?

This impression is indeed misleading. Mexico is more densely populated than the U.S.A., with 68 acres per person compared with the U.S.A.'s 106. Worse still, Mexico has only seven acres of arable land per head while the U.S.A. has 21. Moreover, account must be taken of land quality. In a rich country such as the United States, there are many millions of acres of marginal land which farmers do not cultivate because even by applying the most advanced technology they would not yield an acceptable income. In Mexico, where income expectations are lower, much marginal land is under the plough and grows crops of corn and beans so meager that only subsistence farmers with no alternative means of livelihood operate them. As the economic situation improves, and job opportunities increase, some of this land is beginning to be abandoned, and during the eighties more of it will go out of cultivation.

Nevertheless, continues the argument of the illusioned, there exist 15 to 20 million acres of virgin land in the southeast, flat or gently undulating plains in a zone of good rainfall which could become the food basket of the future. Because of its potential significance we devoted some space in Chapter 3 to examining this contention and reached the conclusion that this is not merely an illusion but a dangerous one because, by entering that region with bulldozers and scrapers, the fertility of those soils can be destroyed and the land made sterile for generations. This has occurred in similar ecological circumstances in other countries and indeed in the few areas in Mexico's own southeast where in the 1970s an attempt was made to carve large new ejidos out of the tropical bush.

However, our illusionists are not yet vanquished. They still have one more card to play, which is water. Calculations have been made of the number of cusecs (cubic inches per second) of water which flow down Mexico's rivers, and when the resultant figure is compared with the amount of water used in irrigation and for human and industrial purposes, it is found that the percentage of utilization is quite low. So why not harness more water and irrigate the arid areas? Unfortunately, the unused water supplies are located in the wrong places. Already, for the consumption of Mexico City, water is being brought over great distances at great expense. In respect to farms, it was the general conclusion of the National Water Plan that the irrigation needs lie chiefly in the north and center, whereas the big untapped rivers are located in the south and southeast. In the north the only major project still on the drawing board is an extremely costly one which would bring water from the state of Nayarit to the northern part of Sinaloa and to Sonora, a distance of some 500 miles. In the south the Grijalva and the Usumacinta, to mention only the two largest rivers, pour billions of gallons every day into the sea; yet not only is their topography difficult for constructing irrigation works (though hydroelectric dams are already being built), but the land which the water would serve lies in a zone where cyclones may strike more than once each season and where the major preoccupation is not drought but flooding, which may persist for several months in succession. Even the water engineers who drafted the plan and who tend to optimism in matters touching their profession, concluded that irrigating the southeast must, for a long while to come, remain a gleam in the eye.

A rather different version of the spatial illusion appears in the contention that the existing areas under cultivation used more rationally could achieve national self-sufficiency in food supplies. The proponents of this thesis argue that food self-sufficiency is a prerequisite of economic independence, that it would be politically dangerous to utilize part of the plethora of petroleum to finance an ever increasing volume of food imports. Adherents to this school of thought cite the Uncle Sam bogey, alleging that such a policy would merely recreate the dependence from which Mexico has been trying to escape for the past several decades, that the U.S. grain export embargo of 1980 showed once again that food can be used as a political weapon.

In the summer of 1980 the Mexican government launched, with much publicity, an agricultural plan entitled *Sistema Alimentario Mexicano*, which forthwith was nicknamed SAM. This was described as the López Portillo administration's principal and definitive contribution to solving the problems of the farm sector. Its specific goals were to achieve self-sufficiency in corn and beans by 1982, and in wheat, rice, sorghums, and oilseeds by 1984. It also contained vaguely worded proposals for improving the nutrition of 19 million malnourished Mexicans.

The measures outlined for attaining the agricultural objectives included: (1) an increase of 30 percent in the guaranteed price of corn, and of 50 percent in that of beans; (2) a reduction from 14 to 3 percent in the interest charges on production credit for these two crops; (3) a reduction of 75 percent in the prices of seed corn and bean seed, and of 30 percent in the prices of fertilizers used on those crops; (4) a reduction to 3 percent in the crop insurance premiums for these commodities; and (5) the expropriation of 3-4 million acres of ranch grasslands for transfer to ejidatarios to sow corn and beans.

At this writing, it is too early to say how far this plan will be implemented in practice. There have been many previous agricultural plans whose texts have accumulated dust in the government's archives. If SAM is really pursued with vigor, it will represent a triumph of the demogogues, who preach self-sufficiency in "basic feedstuffs," over the economists mindful of comparative costs who advocate the importation of animal feeds, so as to achieve a greater output of milk and meat, thereby giving more employment and improving the levels of human nutrition.

For example, if parts of ranches are expropriated, these will probably consist of the better quality pastures, since only on these would there be any chance (and that rather dubious) of producing satisfactory quantities of corn and beans. For the ranchers the consequential choice would be: either over-stocking and further deteriorating their remaining grasslands or slaughtering parts of their herds.

The promised reductions in the prices of seeds and fertilizers, and in interest on crop credit, will entail substantial subsidies, as will also the increases in the guaranteed prices, unless these are to be reflected in higher prices at retail level. (And these guaranteed prices assist the larger-scale producers, not the poor mini-farmers.) Similarly, a program for improving the nutrition of many millions of poor families will, depending on its scope, entail a major call on public funds. Considered from any angle, the measures will aggravate the inflation.

While SAM may possess emotional appeal, its economic foundations are weak. The spatial illusionists need to alter their angle of vision, and concern themselves more with the advocacy of policies which would on the one hand improve the income position of small farmers, particularly ejidatarios, and on the other would contribute to raising the nutritional level of Mexican children. As has been argued in previous pages, this would mean elaborating a livestock production policy that would increase at acceptable prices the supply of animal protein foods on sale to consumers and would be so structured that the major part of the increase in production would come not from large ranches but from small farms.

Such a new orientation would, of course, be incompatible with any goal of self-sufficiency or with clinging to the traditional corn/bean syndrome, since it would require accepting an expansion of imports of animal feed. Because these can be had on the world market at prices lower than those paid to Mexican farmers, the increased output of meat, milk, and eggs would be obtained at lower average costs, and would be within the reach of the poorer consumers.

To enable small farmers to participate in the expansion, a special campaign would be necessary, orchestrated jointly by the extension service and the farm credit institutions, to teach small farmers how to utilize their limited acres to best advantage for feed crops and to finance their purchases of feed mixes. A development of this kind was successfully accomplished decades ago in such countries of small farmers as Denmark and the Netherlands. It is a task to which the recently proposed cooperatives, joint ventures between private enterprise and ejidatarios, could address themselves.

REVOLUTION WITHOUT END

In the discussion of Mexico's agrarian reform we had occasion to note how it had been an integral part of the Revolution of 1910 and how frequently it is claimed that, philosophically speaking, the Revolution is still going on. This doctrine of continuing revolution which sounds more Maoist than Latin has divers explanations. It is maintained by some who feel that the original goals of the Revolution, in respect to egalitarianism or socialism or some other ideal, have not yet been fully achieved and that therefore the struggle should continue. It is maintained by those who feel, like the Maoists, that the governing class and the bureaucracy have become too set in their ways and should be from time to time overthrown. Finally, it is maintained by more cynical leaders who wish to preserve the stability of existing institutions and who see wisdom in talking revolution in order to defuse any agitation in favor of change.

For indeed the talk is one thing, and, as so often, the practice is another. Mexico's institutions, political, social, and economic, are in no sense in a state of flux, but are remarkably set in their form and their ways. The country possesses a pattern of politics—the one party state and the manipulation of aspiring individuals—which in its consistency and stability could be characterized as extremely conservative. It possesses an agrarian pattern which, because it was established by the original revolutionaries, has become hallowed and revered although partly out of date. Perhaps a nation, which in the more distant past experienced so many political upheavals, only naturally becomes conservative when it finds itself with institutions which, at any rate for a time, have worked tolerably well.

However, one awkward consequence of combining revolutionary talk with conservative practice, is the absence of any public discussion of alternatives. No journals exist in which one might find articles canvassing alternatives or modifications to the present political arrangements; none which examine dispassionately the pros and cons of, for instance, more or less State intervention in the economic sphere (except for the avowedly Marxist publications which preach total Statism). In respect to the farm sector there is no forum for discussing openly and publicly how the institutions of the Agrarian Reform might be adapted to serve better the situation in the eighties which is so different from that of seventy years ago.

Naturally, a certain amount of verbal discussion takes place at the so-called "political breakfasts" frequented by leaders of the various pressure groups, and occasionally a few sentences of suggestions may be incorporated in a political speech, but a rational consideration of the disadvantages of the existing land tenure system is taboo. This strange vacuum arises partly from the above-mentioned imbalance in the party structure, with vociferous opposition to the left of the PRI and total silence to the right. In terms of agrarian policy the choice is therefore limited to agricultural collectivization or going forward as we are. No body of opinion expounds any coherent third point of view.

In 1980 it was not yet possible to admit publicly, though it was widely accepted privately, that the ejido as an institution had failed, inasmuch as it had never performed the services of cooperative purchase and marketing which were envisaged, and that the farm tenure provided for ejidatarios had not mobilized whatever production initiative they might have.

Inasmuch as there is little public discussion of the validity or otherwise of this diagnosis, there is even less about what could be done to amend in a helpful way the existing agrarian legislation. We have suggested that ejidatarios should be granted titles to their land, not titles of absolute ownership but rather usufruct. They should have the right to rent out and, subject to certain qualifications, sell their parcels. If they decide to quit the ejido, they should receive compensation for capital improvements.

In regard to private sector farmers, we have urged that they too need titles which should be similar to those proposed for ejidatarios, i.e., thirty-year rent-free leases from the State, renewable at the operators' option. We have proposed that the limits to size of farm be retained, but that the legal texts need rewriting to remove their grave ambiguities.

With the legislation adequately redrafted and with land titles granted to all operators complying therewith, the legality of the occupancy would have been officially recognized and confirmed; thereafter any invasion by squatters or would-be ejidatarios would rank as an indictable offense to be dealt with by the courts. Similarly, the ejidatarios' contracts would protect them against arbitrary action by ejido presidents and local caciques.

A packet of measures on these lines would give new life to the institutions of the Agrarian Reform. In this way it could indeed be claimed that the Agrarian Reform remains a continuing principle in the fabric of rural society. Just as all Mexicans wish to be considered, in some sense or other, good revolutionaries, so all of them could claim to be good and loyal reformers.

THE STRAIGHTJACKET OF BUREAUCRACY

Many references have been made in previous pages to the built-in tendency of Mexican bureaucracy to multiply itself; not that this does not feature prominently in the bureaucracies of other countries, but rather that with the monolithic political structure of Mexico it is not possible for counter-vailing movements to develop as they have done in, for instance, the U.S.A. Indeed, our scenario for the eighties, with a steady increase in the dominance of the public sector, presupposes that bureaucracy will increase in parallel, because it is the unsaid presumption that the State is the number one problem-solver.

In the farm sector this attitude of "everything should be handled by government" finds expression in several fields. It is taken for granted that the opening up of new land for cultivation must be a government project and that any construction of new irrigation facilities must be a public works activity. It is considered unrealistic to propose that village people should be capable of carrying out on their own initiative local projects for bettering their incomes and the amenities of their villages. It is accepted without question that to allow the prices of farm products to be determined by the market would be catastrophic and that the only way to deal with shortages is to make the State price controls ever more rigid.

There is not much chance of these attitudes and beliefs being substantially modified during the eighties, and yet if they could be made only a little more flexible, immense forces for agricultural progress could be released. Consider the task of land reclamation and settlement, for example, which when undertaken by a government department requires the deployment of vast investment resources because of the project's complex organizational structure, the quantities of costly machinery and equipment, the overmanning and the elaborate overheads, not to mention the many financial leakages. Even with petroleum revenue coming in, the government will have too many other investment commitments to be able to dedicate large sums to these purposes.

Yet this is preeminently a field in which individual efforts and initiative are appropriate. Give a Mexican peasant a land grant of 50 acres and he will be off with his machete to clear his plot, build a home of sorts and establish himself—at no cost whatsoever to the public purse. Because he cannot afford a bulldozer he will not wreck the fertility of his soil; and because he will sow only a few acres of any one crop he can defend himself more effectively against the onslaught of pests and diseases.

Very similar consideration applies to the future development of irrigation in Mexico. The grandiose projects lying in government offices, awaiting allocation of funds, will, for the above-mentioned reasons, be immensely costly to execute and therefore will only be undertaken slowly by installments. Yet, by activating the permission clauses in the National Water Law and granting permits to individuals to exploit at their own expense surface streams and groundwater resources (always within the maximum limits of irrigation acreage laid down in the Reform Act), a very substantial addition could be made to the crop growing potential of the country. The work would impose no financial burden on the State, it would be done economically and efficiently, and it would be located in districts with high chances of economic viability, since the initiators would be certain to investigate carefully before investing their own money.

Also in the realm of farm credit, much greater flexibility is called for. The Rural Credit Bank has devoted its resources excessively to short-term crop credit, partly because the demand for long-term improvement credit remained weak so long as neither ejidatarios nor private farmers enjoyed sufficient security of tenure to make such investments attractive.

With the modifications here suggested in tenure arrangements and the generalized distribution of land titles, this demand would become active. Loans would be sought for land terracing, draining, orchard planting, small irrigation works (under the new dispensation), land colonization, livestock purchase (under the new project for small farms) and so on. The private banks could take an active part alongside the Rural Credit Bank, since their clients would have titles to offer as surety. And with our proposed phasing out of interest-rate subsidies the two components of the banking sector would be competing on equal terms.

As for the prices of farm products, the labyrinth of bureaucratic machinery for their fixing and for the administration of subsequent controls has every prospect of becoming more complex during the present decade. Hardly a week passes without a press announcement that the authorities have begun examining the case for an increase in the retail price of some product; whereupon that article promptly disappears from the stores until a decision has been reached and a new price decree issued. As we have insisted, this policy discourages production and provokes scarcities. Furthermore, no market can be straightjacketed by fiat for very long. Sooner or later the price fixers have to bow to the inevitable and accept the verdict of supply and demand. Progressive decontrol of retail prices would be a blessing for consumers in the long run.

With respect to the regime of guaranteed farm gate prices, our suggestion, that prices of certain important products be differentiated according to each operator's volume of sales, is based on the belief that such an arrangement would be compatible with maintaining production incentives (provided the scale of differentials were prudently selected) and at the same time would provide some tangible relief at the poverty end of the farm community. Such an arrangement should be regarded as a transitory measure, suited to the next few years while the other proposed reforms are accelerating the modernization of the sector. At some later time, when the land tenure system has been improved and when commercial farming has spread to most of the areas where at present subsistence agriculture still prevails, the pricing system should be reconsidered in the light of the then situation.

There are a number of other proposals made in previous chapters which will involve a loosening of the controls imposed on agriculture by the authorities, but they need not be recapitulated here. In farming, as in other sectors of the Mexican economy, there persists too much government, and excessive rigidity in the administration of controls. The bureaucrats who operate the system have a vested interest in maintaining and extending it. Only a gradual stirring of public opposition, together with a grudging recognition that the system functions badly, can ultimately bring into existence something more flexible and more stimulating.

BENIGN NEGLECT — OR?

Nevertheless, it would be wrong to convey an impression of generalized despondency. The 1980s are almost certain to bring more prosperity to Mexico than did the seventies. The boom of petroleum production will strengthen the economy as never before, and will provide new opportunities which hopefully will be seized for industrial diversification and for a more rapid development of exports of manufactures. Increased investment resources will be generated internally, and simultaneously Mexico will become an attractive country for lending and investment by foreign banks and corporations.

It is true that we must expect a persistent trend toward more Statism and more widespread bureaucratic interventions, but these are characteristics also found in many other countries. It is true that paternalism will continue to dominate the relationship between government and people as it has done for centuries. Yet the private sector will soldier on quite successfully in some branches of industry and still more so in commerce, tourism, and other services. Many enterprises have developed techniques of living with inflation, which they discount in advance and thus maintain their profits. Those which hitherto were heavily involved in producing articles subject to price controls, will be trying to diversify away toward articles whose markets remain free.

Relatively high rates of inflation will persist so long as the authorities insist on combining high levels of investment and of government expenditure with a policy of deliberately maintaining and indeed improving the purchasing power of the workers. In the longer run this continuing inflation will make it progressively harder to maintain unaltered the peso's exchange rate, though the government will be reluctant to permit even modest readjustments unless and until it is forced to act. This exchange disequilibrium sooner or later will affect the volume of earnings from tourism and make it more difficult for Mexican firms to sell their products in foreign markets.

Continuing inflation will also complicate the orchestration of industrial expansion and could in consequence retard the rate of growth of the national product. What has occurred is a fall in the volume of private savings, leaving the banks with diminished resources for onward lending. And of these diminished resources the Central Bank commandeers a larger proportion to help finance the public sector enterprises which do not sufficiently generate their own reinvestment resources. Yet the banks will contrive to live with this situation, just as firms will adjust themselves to the likely continuation of high interest rates.

Thus, all in all, in spite of difficulties which may prevent a full realization of the hopes stimulated by the oil bonanza, the economic barometer seems set at "fair" for the next several years. In such circumstances the authorities may well prefer not to take any far-reaching initiatives on the agricultural front, which might have politically destabilizing consequences. If such a policy of benign neglect is maintained, as it has been during the Lopez Portillo administration, then certainly agricultural production will continue to increase only slowly and the need for food imports will become that much greater. But such imports, even in larger volume than at the beginning of the eighties, can be paid for by using only a fraction of the foreign exchange earnings generated from petroleum exports. From a balance of payments point of view there will be no problem.

Furthermore, the incidence of rural poverty will be becoming less severe. A continuing and probably accelerating migration to the cities (and in part to the U.S.A.) will relieve the demographic pressure in rural areas, stabilizing the farm population before it begins its destined decline in absolute numbers. At the same time the government's determined efforts to decentralize industry and other activities will be beginning to bear fruit providing more employment in locations accessible to underoccupied farm families. Though rural poverty will continue to exist, especially in geographically remote areas difficult to reach through aid programs, the afflicted each year will be fewer in number.

As to the Agrarian Reform, the more far-sighted political leaders are aware that it needs to be adapted to the requirements of the present day. They know well the reasons which are holding back agricultural expansion both in the ejidos and the private sector. It has been their judgment up till now that to do more than tinker with the fringes of the problem would open a Pandora's box of social tension and political protest, even leading to local outbreaks of violence. The entrenched ideology of the old Agrarian Reform has such widespread ramifications in the thought structure of Mexican society that attempts to modify it substantially might stir up movements directed toward changing other even more fundamental elements in the country's political institutions.

With preoccupations such as these, it is only natural that the leadership opts for leaving things more or less as they are in the farm sector; better a policy of benign neglect with negative consequences which the nation can comfortably tolerate in its petroleum era, than one of overeager restructuring which, while favorable to agriculture, might provoke undesired instability in the social and political fabric.

And yet we are perhaps entitled to ask, is this necessarily the last word? Would it not be possible to envisage some alternative scenario emerging, even before the end of the decade? Perhaps petroleum prosperity may not merely oil the creaking machinery of government but also, as it becomes more widely diffused, promote a more relaxed attitude to the established social and political institutions, such that it becomes possible to contemplate change without disruption, reform without revolution.

In that event, new vistas would open up for the farm sector. In such a political climate it would become feasible to go forward, step by step, with a reshaping and modernizing of the Agrarian Reform institutions along the lines already sketched out in Chapter 8 so that farm operators, ejidatarios, communeros, and private farmers alike, enjoyed the security and the incentives necessary for achieving real technological progress. The constraints on land colonization and water exploitation by individuals would be progressively removed, thereby shifting much of the investment burden from the government to the private sector. At the same time, a restructuring of farm product prices to favor the smaller producers and to encourage family enterprises in livestock, together with a reorientation of farm credit policies toward long-term improvement objectives would assist the sector in adapting more flexibly to the changing pattern of consumer demand and would powerfully contribute to improving the living standards of the lower-income cultivators.

The latent potential for further agricultural expansion in Mexico remains great. Per acre yields can be substantially raised, especially in the large middle zone which lies between the high technology irrigation districts on the one side and the marginal semiarid lands on the other. Livestock can be increased in numbers and in quality, once the agronomic and economic problems of the feed supply have been resolved and the interest of small-scale operators has been aroused. If these opportunities were seized there is no reason why the present lethargic growth of farm output at between 2.5 and 3.5 percent per year should not be transformed into a rate of 4 to 5 percent as it was in the golden age of Mexican agriculture.

Probably the next few years are going to be the most difficult. Population and demand will be growing rapidly while new agricultural policies will not yet have been formulated and implemented. Further ahead, the skies begin to brighten. Population increase will be slackening and demand expansion less dramatic; the nettles of agrarian and economic reform will become easier to grasp while decentralized industrial expansion, stimulated by petroleum, will mitigate the problems of un- and underemployment.

We might visualize a gradual elimination of the social cleavage between ejidatarios and private farmers as the institutions governing their activities become progressively more homogenized. Distinctions would not be entirely obliterated. On the one hand there would be the larger farmers, in crops and livestock and in mixed farming, possessing more skills and commanding more resources; while on the other there would come into existence for the first time a class of small-scale yet prosperous family farmers, most of them producing both crops and animal products in various combinations. With the gradual reduction in population pressure these family farms would not become fragmented, but at the other end of the scale the legislation would prevent farms from becoming too large. Thus the economic as well as social integration of Mexico's farmers into a single community should be well on its way.

Mexico indeed faces an agricultural dilemma at the present time, because her agricultural needs and her agrarian institutions are at cross purposes. Perhaps, unfortunately, it is a dilemma easily shied away from, because the blessing of oil will permit the postponement of decision-making on the farm front. But such postponement, if persisted in, would bring about grave distortions in the nation's economic balance, leaving its agricultural resources underutilized and its farm people in continuing poverty. Yet with the rapid accumulation of economic abundance and with reaching political maturity, facing the dilemma squarely is likely to be less difficult than many people imagine. The problems are not all that insoluble, the solutions not all that radical. The prize would be of prodigious proportions: nothing less than the emergence of a united, progressive, and prosperous farming community, providing efficiently the varied nutritional needs of a nation destined to play an ever more important role in world affairs.

Appendix: Sources of Data

LAND USE

The principal source of statistical data on land use is the decennial agricultural census first published by the Statistical Office in 1930 and giving, in its General Summary, particulars for each of the 32 states of the Republic. Separate volumes, one for each of the states, give details by *municipios* (equivalent to U.S. counties). In addition, the annual reports of the Agricultural Economics Branch of the Department of Agriculture provide data on the *harvested* area of each of the more important crops.

In attempting to identify historical trends in the census material, one encounters various major anomalies, but for the purposes of this study only those relating to the arable area have been examined, because this is such a basic concept that, unless the worst errors could be removed, any analysis of past developments and future potential would be impossible. The official definition of arable is clear, and has not been changed; it refers to any land which has been cultivated at least once during the previous five years, whether or not it is in cultivation at the date of the census.

The census errors were chiefly of two kinds. The first concerns the treatment of the practice of shifting (nomadic) cultivation, common in the southern and peninsular states, where the operators burn a patch of forest, cultivate for two or three years, abandon it and move on to another patch. Frequently the enumerators recorded as "arable" any land which has been cultivated within living memory. As a result the fallow in many cases came to 90 percent of the arable area, instead of the 30 to 40 percent more normally found in other parts of the country. This error was largely corrected in the 1970 census causing massive apparent declines in the arable areas of some southern states.

The second error was the generalized tendency in 1960 to overestimate all areas, not only arable but also other land use classes, this being especially notable in Chihuahua, Coahuila and Durango. In Campeche the census reported 20 million acres of "land in farms," whereas the entire surface of that state totals less than 13 million acres.

The principles followed in attempting corrections to the reported arable areas have been to consider state by state: (1) the decade-to-decade changes in the areas under perennial and under annual crops, each of which is reported separately, and (2) what would be a normal proportion of fallow in the state being examined. The area under the two types of crops plus a reasonable area of fallow is presumed to add up to the arable area of that state. It was further assumed that there should not occur too violent fluctuations in a state's arable area from census to census, especially where there was present an obvious long-term upward trend, as was generally the case. In this way it became possible to establish consistent historical series for arable and for fallow. Corrections of less than 250,000 acres in any one state have usually been ignored.

A particular anomaly arose in the 1970 census where a new category of land use was included, namely "cultivated grasslands." This was intended to cover areas sown to perennial grasses, such as sudan grass, pangola, star of Africa, etc. In practice the enumerators in many states also included ordinary pastures which had been periodically cleared of bushes and stones. In any case this land, which in 1970 totaled 10 million acres, does not carry crops and more properly belongs to "pasture"; accordingly in this study it has been transferred from the arable to the pasture classification.

PERENNIAL CROPS

Again the two sources of information are the census for planted area and the Department of Agriculture for harvested area. However, all the censuses have included as plantation perennials the various types of maguey, whereas the latter source omits these items.

Also, over the years there have been changes in respect to what are classified as annual and what as perennial crops, for instance in the cases of coffee, henequen, alfalfa, and sugarcane. To establish consistency through time we have used the division adopted in the 1970 census and have adjusted the earlier censuses and the Department of Agriculture's annual reports to conform therewith.

THE 1970 CENSUS

In this census for the first time the authorities attempted to record separately the data for the winter crop cycle and for the spring/summer cycle. This caused great confusion, for not only do the two cycles differ in time span in different parts of the country, but of course in any single district they overlap. The result was a good deal of double counting (the same crop area being recorded in both cycles), an exaggeration of some 4 million acres in the total sown area, the almost total elimination of fallow in some states (e.g., down to 0.2 percent in Veracruz) and a doubling of the area of crop losses.

An examination state by state made it possible for the introduction of certain corrections in the present study, sufficient to arrive at credible figures for the gross sown area, for the crop losses in that year, and therefore for the harvested area. But since the 1970 census gave no information on the area double-cropped, we could not ascertain the net sown area nor, consequently, the fallow.

THE HARVESTED AREA

In the earlier years of crop reporting there was a reasonable degree of concordance between the farm census and the Department of Agriculture, and this persisted up to and including 1960 (although some authors alleged that in the late forties and early fifties the department was underestimating the area and production of corn and beans). But by the time of the 1970 census (the data referred to land use in 1969), the two sources were far apart in regard to harvested area, the census giving a figure of 26.1 million acres and the department one of 32.3 million. Especially large differences emerged for corn, beans, grain sorghums, and soybeans. In certain states the Department of Agriculture recorded a total harvested area more than twice as large as that reported by the census in the same year: in Jalisco, Querétaro, and Zacatecas, for instance. Consequently, the two sources published widely different figures of the volume of production.

CROP YIELDS

Yields per acre for each reported crop are published in the census, but since these refer to a single year with its climatic peculiarities whatever they might be, this does not tell us much. Hence the sole source for historical series is the Department of Agriculture. Incidentally, in the individual years when both sources report, their differences of opinion may be substantial: there are cases where the national average per-acre yield of a particular crop according to the department was twice as high as that reported by the census.

More disturbing, however, are the changes recorded *within* a particular series as published by the department. Normally we expect to find a long-term upward trend in per-acre yield, reflecting advances in technology, a trend which might average around 2 percent per year, but higher or lower than this according to crop. Yet in many crops the published figures show sudden jumps where the yield rises 50 percent, or in some instances even doubles, from one year to the next, thereafter resuming its slow upward improvement. Notable examples, among many, are potatoes, tomatoes, avocados, mangoes, and strawberries. Apparently at a certain moment the authorities concluded that their reporting procedures had been inaccurate, and so introduced changes. Since in almost all cases the revisions have been upward ones, the consequence is to present an exaggerated picture of the long-term rate of improvement in yields. No adjustments are feasible in this matter, but we need to be aware of the order of magnitude of the distortion.

In order to obtain guidance, we compared over a period of twenty years the improvement in Mexican yields with those of other major Latin American countries in respect to corn, beans, barley, wheat, cotton, and potatoes. During that period the yield improvement in Mexico was respectively 31 percent, 111 percent, 110 percent, 156 percent, 189 percent, and 54 percent greater than the average gain of the other countries. This is all very flattering to Mexico, and while we may be prepared to believe that Mexican farmers have been *somewhat* more progressive than their South American colleagues, it is hard to accept that they have been all that much more successful.

CROP PRODUCTION

Since the two sources diverge markedly in regard both to crop areas and crop yields, the department figures in both matters being higher than those of the census, it follows that their differences are still larger when it comes to reporting the volume of production. Thus between 1950 and 1970 the one reports a volume increase of 130 percent and the other one of 180 percent.

Although this is a delicate matter to mention, it would appear that in Mexico some politicization of statistics has been taking place. Because the government has publicly committed itself to modernizing agriculture, expanding production, and rapidly achieving self-sufficiency in food supplies, a junior official at county level realizes that his immediate superior will be displeased if he produces crop figures worse than those of the previous year, and will be delighted if they are better; so he may make certain "adjustments." The situation repeats itself when the county reports to the state capital, when the state submits its data to the Department of Agriculture and when its officials correlate the information for presentation to the undersecretary. There was one occasion in the late seventies when the Secretary of Agriculture took the undersecretary's figure for the corn harvest and added 30 percent before making his announcement.

On the other hand, the census enumerators, being ad hoc temporary employees, have no careers at stake and no built-in bias. They have their shortcomings, indeed many, but not this one. Probably, therefore, the census statistics are in general more credible than those of the Department of Agriculture, and in this study we have preferred using them wherever possible. But for data subsequent to the 1970 census our only source is the Department of Agriculture, since the results of the 1981 farm census (it is being taken one year later) are not available at the time of this writing. Meanwhile the possible defects of the figures for the seventies should be borne in mind.

THE IRRIGATION DISTRICTS

The administration of these districts was the responsibility of the Federal Water Resources Department until 1977 when it was merged with the Department of Agriculture. The Water Resources Department published annual reports, which are still being continued, giving data on the area, yield, output, and value of output of each crop in each irrigation district, as well as separate figures for the output of ejidatarios and of private farmers. In addition occasional reports are issued on special topics such as livestock activities in irrigation districts, degree of mechanization, production costs of selected crops, etc. The reports maintain high professional standards.

Nevertheless, care is required in the interpretation of some of the concepts. For instance, "irrigated area" is a vague term which may mean different things in different contexts. Sometimes it refers to the total area of land within the perimeters of the irrigation district, including roads and buildings and open spaces as well as agricultural land. Sometimes it indicates the area of land actually being farmed in the district but includes areas which, because of undulations etc., cannot ever be irrigated. Or it may denote the area which at one time or another was irrigated, although currently some portions may be out of use through soil salination, silting up of canals, etc. Often the phrase fails to distinguish between net irrigated area (the acres to which water was delivered in the course of the year) and gross irrigated area (which includes the above plus the areas on which second crops were grown and which therefore received further water deliveries). Careful reading of the reports usually indicates in which of these various senses the phrase ''irrigated area'' is being used.

A more intractable problem arises from the contradictions that can be noted between these irrigation district statistics and those of the Department of Agriculture which purport to refer to all the irrigated lands of the Republic, whether inside or outside the so-called irrigation districts. For example, in some instances the Department of Agriculture's figure for the entire irrigated acreage of a certain crop may be lower than that for the irrigation districts, though the latter physically form only part of the national total. Conflicts also exist with regard to per-acre yields and the volume and value of outputs.

There have also been discrepancies between the irrigation districts and the farm census. Thus in 1970 the irrigation districts reported that the harvested area constituted 111 percent of the irrigated area, reflecting a certain amount of double cropping, whereas the census reported only 84 percent. If the census were correct, we would have to believe that, of the irrigated areas outside the districts, only 45 percent of the land bore harvested crops, a most unlikely situation. In this case it is the census which would appear to be at fault.

LIVESTOCK NUMBERS

The most notorious contradictions between the farm censuses and the Department of Agriculture's annual reports relate to numbers of livestock. In 1970 the department's figures surpassed those of the census by 2.7 million cattle, 1.5 million horses, 1.4 million mules, 1.1 million hogs, and 3 million sheep. They fell short of the census in poultry by 27 million. In each of the states of Chihuahua and Veracruz the department in that year counted over one million head of cattle more than the census, in Zacatecas 800,000 more sheep, in Veracruz 445,000 more horses, and so on.

We cannot assert that the census figures are correct; undoubtedly they contain errors of their own. But there are at least two reasons for preferring the census data. First, the Department of Agriculture's statistics of livestock, like those of crops, tend to be politicized. Secondly, the department has a poor record for internal consistency through time of its own series of data. Thus around 1960 it was recording over 30 million head of cattle although in previous and subsequent years it published figures of only 20 to 22 million. In individual states the anomalies were greater. For instance, in San Luis Potosí in 1960 only 15 asses were reported, but in 1965 there were 97,638. Sometimes it would appear that a decimal point slipped: Coahuila in 1960 registered 1,148,785 hogs, but in 1959 and again in 1961 around 150,000.

During the seventies the Department of Agriculture made an effort to improve its reporting procedures, with individual figures being carefully reviewed prior to publication. This resulted in removing most of the arithmetical errors, but no means were available for dealing with the politicizing factor. The arbitrary "smoothing" operation may even have gone too far, since it seems to have eliminated the well-known hog cycle which Mexico, like other countries, previously experienced.

LIVESTOCK OUTPUT

The censuses have always published figures of the output of milk, cheese, butter, eggs, wool, honey, and beeswax, but never of meat. The Department of Agriculture's publications include these items and also all classes of meat. In milk the department's figures have been lower and for eggs considerably higher than those of the censuses. A number of other agencies and authors (in monographs) have put forward figures of their own. The confusion has been so great that the present author felt obliged to reexamine the whole topic.

In regard to meat, most Mexican experts agreed that the output was underestimated; the question was—by how much? As a first step we established comparisons between Mexico and other Latin American countries in respect to (1) per capita meat consumption, and (2) the alleged meat output per thousand head of cattle and of other classes of animals. It at once emerged that on both counts the Mexicans ranked almost the lowest in the whole continent, which does not accord with practical experience.

So, taking into account differences in per capita income and food consumption habits, plus whatever differences between the countries were known to exist with regard to herd quality, new figures were sought which harmonized better with those in countries whose situations were similar. The result was an annual meat output of around 2 million metric tons in the mid-seventies, implying a per capita consumption of 68 to 70 pounds, which was more than double the quantity being reported by the Department of Agriculture, but which left Mexico still a little below the Latin American average. It so happened that the National Institute of Nutrition was publishing for the same period figures very similar to ours.

It may be mentioned that, as in livestock numbers, the Department of Agriculture during the seventies claimed to publish more reliable annual statistics of meat production, yet these contained some strange anomalies. For example, according to the new series, betweeen 1970 and 1978 the meat output per thousand head of hogs increased 50 percent and per thousand head of cattle by 26 percent and per thousand head of poultry by 175 percent. Such large productivity improvements are unlikely to have been achieved within the short span of eight years.

In regard to milk, the census data for 1970 gave an output and a per capita consumption both of which appeared plausible according to our tests, and which have been retained, although the department's figure for that year was almost 20 percent less. Again, however, between 1970 and 1978 the department postulated a 33 percent increase in average milk yield per cow, which exceeds all bounds of feasibility.

On the other hand, with regard to eggs the census again appears to have seriously underestimated the volume of production. The data published by the National Nutrition Institute seemed to accord better with the known state of the poultry industry and with indications in monographs on the subject. Thus the institute's figures for the seventies were retained as a base for calculations backwards and forwards.

In order to establish time series for meat, milk, and eggs, the only firm indicators available were those of livestock population in the census series, which we have shown to be preferable to the department's series. But it was necessary in addition to establish assumptions regarding coefficients of productivity per head (or per thousand head). By a process of trial and error the following productivity growth rates were fixed on: for beef cattle, hogs, sheep and goats 1 percent per annum from 1930 to 1980; for dairy cattle 1.5 percent from 1930 to 1960 and then 1 percent from 1960 to 1980; for poultry (meat and eggs) 2 percent throughout. The use of lower growth rates in productivity would have resulted in livestock production far surpassing crop production in value in the thirties and forties, a proposition which no one would believe. The use of higher rates would have produced levels of consumption in those earlier years far too low to be plausible. Admittedly the procedure adopted is far from scientific, but it seems to give results that correspond better than do any of the official figures with what is known of the Mexican situation in general.

While this procedure enabled us to construct a historical series for the country as a whole, we were still unable to measure the progress of the different regions and of the states within those regions. To find a way of making some rational guesses on this theme the following methodology was adopted. From the data in the 1970 farm census the productivity of the different classes of livestock could be calculated by states in terms of value of output per thousand head in each category of animals. (Although the data were underestimates in absolute terms, it had to be assumed that the underestimation was proportionately similar in all parts of the country.) In accordance with these calculations the 32 states were divided into three groups of high, medium, and low animal productivity.

It was then assumed, on the basis of experience with other economic indicators, that twenty years earlier the differences had been less pronounced. In other words, between 1950 and 1970 certain states had modernized their production more than others, and those reaching high productivity levels in 1970 had advanced more rapidly than those recording low productivity levels in that year. For each of the three groups of states coefficients of productivity increase over the twenty-year period were selected to mirror these differences in rate of progress: 40 percent, 25 percent, and 15 percent respectively. Checking backward revealed that the weighted average of these calculations produced figures of national livestock output in 1950 close to those mentioned in previous paragraphs.

This methodology was also used for the calculation of the livestock element in total farm output (crops plus stock) by regions as elaborated in Chapter 6. Further, in order to express this output in per capita terms for 1970, there remained the problem of the widely differing estimates of farm population for that year, and our own conviction that the population indeed fluctuates between, say, 3 and 10 million from one season to another. It so happened, however, that the figure emerging from the population census of 1970, namely 5.1 million, seemed to represent a reasonable midpoint between the extremes just mentioned. Accordingly the state by state farm population data from that census were utilized for our per capita calculations.

FARM SIZES

This is a topic on which unfortunately the statistics have little to say. In each and all of the five agricultural censuses the main tables are broken down as follows: (1) private farms of more than 12.5 acres, (2) private farms below 12.5 acres and (3) ejidos. This tells us little indeed. Only in respect to three items—total area, arable area, and irrigation area—do the censuses publish a more detailed breakdown, by ten different size classes; and even here the ejidos are treated as whole units though few of them are collectivized. The censuses offered no cross tabulations showing crops and farm sizes or livestock and sizes.

As for the ejidatarios, the censuses prior to 1970 merely published the numbers of operators having parcelas of 0-1 hectares, 1-4 ha, 4-10 ha, etc. and tiresomely these size groups did not correspond with those used for private farms. The parcelas divided at 4 ha. but the private farms at 5 ha; again the former at 20 but the latter at 25 ha.

The 1970 census published two special volumes dedicated to ejidos and communities, to ejidatarios and comuneros. These gave detailed information by finely differentiated size groups (at one hectare intervals up to ten) for land use and for crop output, but not one word was published about these people's livestock activities.

These various shortcomings have created difficulties in presenting a coherent account of what has been happening with respect to farm sizes in Mexico in recent decades.

LAND TENURE

Mention has just been made of the unsatisfactory treatment of ejidos and ejidatarios in the censuses, in spite of the fact that they account for nearly half the nation's farm land and three-quarters of its farm population. Moreover, there has been inconsistent treatment of the *comunidades* which occupy more than 22 million acres. In the first four censuses these communities were included in the totals for private farms of over 12.5 acres and not shown separately. In 1970 they were included with ejidos in all the tables in the General Summary. Therefore, for historical series appropriate adjustments had to be made, which was difficult since it is hardly possible to ascertain the land use pattern in the communities prior to 1970. These adjustments affect land use, crop production, livestock, farm expenses, and other items.

In regard to livestock numbers the censuses reported animals as distributed among four types of farm: (1) private farms of more than 12.5 acres, (2) private farms below 12.5 acres, (3) ejidos (in 1970 with communities), and (4) animals in "populated areas," i.e. in villages. The owners of the animals in this last category might be private farmers, ejidatarios or comuneros. We do not know. Moreover, since in Mexico the great majority of farm operators live, not on their land, but in villages or even towns, where they customarily keep their animals in or adjacent to their homes, the number of animals in the "populated areas" category was huge.

For analytical purposes it was desirable to try to determine how these animals should be distributed between the different groups of farmers. After gathering the meager evidence available, and seeking solutions which would conform with the long-term trends in each group, the conclusions reached were that virtually none of these animals belonged to the larger private farms, and that they should be divided between the smaller private farmers and the ejidatarios in a way which gave a larger share to the ejidatarios in the later years. The figures finally chosen for distributing these "populated area" animals were for 1950 25 percent to ejidos and 75 percent to small private farmers; for 1960 50 percent and 50 percent respectively; for 1970 75 percent and 25 percent respectively. Again these allocations are admittedly arbitrary, but they do accord with the steady decline in the number of small-scale private farmers and the increase in the number of ejidatarios.

Glossary

*

The following are English equivalents (used in the text) of Spanish titles of main statistical sources listed in the bibliography.

Agrarian Reform Act	Ley Federal de Reforma Agraria
Agricultural Statistics of Irrigation Districts	Estadística Agrícola en los Distritos de Riego
Annual Crop Reports	Anuario Estadístico de la Producción Agrícola
Annual Livestock Reports	Anuario Estadístico de la Población y Producción Pecuaria
Bureau of Agricultural Economics	Dirección General de Economía Agrícola
Crop and Livestock Prices (usually published in the crop and livestock yearbooks)	Precios Medios Rurales
Department of Agriculture before 1977:	Secretaría de Agricultura y Ganadería
since 1977:	Secretaría de Agricultura y Recursos Hidráulicos
Department of Programming and Budget	Secretaría de Programación y Presupuesto
Department of Water Resources	Secretaría de Recursos Hidráulicos
Economic Commission for Latin America (United Nations)	Comisión Económica para América Latina (de las Naciones Unidas)
Farm Census	Censo Agrícola, Ganadero y Ejidal
National Rural Credit Bank	Banco Nacional de Crédito Rural
Population Census	Censo de Población
Statistical Office	Dirección General de Estadística

*

[280]

Selected Bibliography

ACOSTA PRODINAT, RODRIGO, *La productividad agrícola en México*, 2d ed., México, 1970.

ALDUCIN PRESNO, JOSE LUIS, *Régimen jurídico de la ganadería en México*, Universidad Nacional Autónoma de México, Facultad de Derecho, México, 1966.

ALTIMIR, OSCAR, *La medición de la PEA de México, 1950–1970*, Comisión Económica para América Latina, México 73/15, May 1975.

ANGULO, HUMBERTO G., "Indice de la producción agrícola," *Revista de Economía*, vol. 9, no. 1, México, January 1946.

ASOCIACION DE AGRICULTORES DEL RIO CULIACAN, *Boletines Agrícolas* (various numbers).

ASOCIACION DE AGRICULTORES DEL RIO FUERTE, *Informe de Actividades Desarrolladas* (various yearly issues).

AZIZ, S. (ed.), *Hunger, Politics and Markets*, New York University Press, New York, 1975.

BALLESTEROS PORTA, JUAN, "El problema de la tierra en México," *Rivista di Agricoltura Subtropicale e Tropicale*, anno 68, no. 1-3/4.6, 1974.

BANCO DE MEXICO, S.A., *Características de la agricultura mexicana*, M. Rodríguez Cisneros, et al., México, 1974.

————, *Características y finalidades de los fondos instituidos en relación con la agricultura*, México, 1976.

————, *Comercio Exterior* (monthly).

————, *Cuentas nacionales y acervos de capital* (various volumes).

————, Fideicomisos Instituidos en Relación con la Agricultura (FIRA), *Informe Anual* (various years).

————, *Informe Anual del Banco de México, S.A.* (various years).

————, *La distribución del ingreso en México*, Fondo de Cultura Económica, México, 1974.

————, *Precios* (monthly bulletin).

————, *Proyecciones de la oferta y la demanda de productos agropecuarios en México a 1965, 1970 y 1975*, México, 1966.

————, *Sector Externo* (monthly).

BANCO NACIONAL AGROPECUARIO, Departamento de Estudios Agropecuarios, *El mercado de ganado porcino en México*, México, December 1973.

————, *La Begoña, Distrito de Riego No. 85, Programa de desarrollo*, México, 1973.

[281]

282   *Bibliography*

BANCO NACIONAL DE COMERCIO EXTERIOR, "Estructuras agrarias y agricultura en México," *Comercio Exterior,* vol. 25, no. 5, May, 1975.

BANCO NACIONAL DE CREDITO RURAL, *Costos de producción de once cultivos básicos,* México, 1975.

————, *Plan de operaciones* (various years).

BANCO NACIONAL DE MEXICO, S.A., *Examen de la situación económica de México* (monthly survey; various numbers). (Also in English.)

BARBOSA RAMIREZ, A. RENE, *El problema del empleo y el problema de la alimentación,* Centro de Investigaciones Agrarias, México, 1976.

————, *La ganadería privada y ejidal—un estudio en Tabasco,* Centro de Investigaciones Agrarias, México, 1974.

CENTRO DE ESTUDIOS ECONOMICOS DEL SECTOR PRIVADO. "El problema ocupacional de México," *Síntesis Económicas,* año 1, no. 11, February 1976.

CENTRO DE INVESTIGACIONES AGRARIAS, *Estructura agraria y desarrollo agrícola en México,* 3 vols., México, 1970.

CENTRO INTERNACIONAL DE MEJORAMIENTO DE MAIZ Y TRIGO (CIMMYT), *The Puebla Project: Seven Years of Experience, 1967–73,* El Batán, México, 1974.

CENTRO NACIONAL DE PRODUCTIVIDAD, *Reforma agraria,* Ediciones Productividad, México, 1969.

COLEGIO DE POSTGRADUADOS, CHAPINGO, *Inventario nacional de los recursos dedicados a la investigación científica agropecuaria en México,* vol. 1, *Instituciones o unidades de investigación,* y vol. 2, *Proyectos y personal científico,* Chapingo, México, 1976.

COMISION ECONOMICA PARA AMERICA LATINA (CEPAL), *La industria de la carne de ganado bovino en México: análisis y perspectivas,* Fondo de Cultura Económica, México, 1975.

————, *Situación y evolución de la agricultura y la alimentación,* División Conjunta CEPAL-FAO, Santiago de Chile, March 1976.

COMISION NACIONAL DE FRUTICULTURA, *Investigaciones fisiológicas* (various numbers).

COMISION NACIONAL DE LA INDUSTRIA AZUCARERA, *Estadísticas azucareras,* Unión Nacional de Productores de Azúcar, Informe Anual.

COMISION NACIONAL DE LOS SALARIOS MINIMOS, *Memoria de los trabajos* (various years).

COMITE TECNICO Y DE DISTRIBUCION DE FONDOS DEL FIDEICOMISO DEL FONDO CANDELILLERO, *Programa de rehabilitación y mejoramiento de agostaderos en el estado de Coahuila,* April 1974.

COMPANIA NACIONAL DE SUBSISTENCIAS POPULARES (CONASUPO), *El Correo Campesino* (various numbers).

CONFEDERACION DE ASOCIACIONES AGRICOLAS DEL ESTADO DE SINALOA (CAADES), *Análisis de la situación agrícola de Sinaloa,* Culiacán (various monthly numbers).

CONFEDERACION NACIONAL CAMPESINA (CNC), *Revista del México Agrario* (various numbers).

CONSEJO NACIONAL DE CIENCIA Y TECNOLOGIA (CONACYT), Centro de Ecodesarrollo, *Los campesinos: para qué organizarlos?* (various authors), México, 1976.

CORDEMEX, *Informes Anuales* (various years).

DOVRING, FOLKE, "Reforma agraria y productividad," in Solís, M. Leopoldo (ed.), *La economía mexicana,* Fondo de Cultura Económica, México, 1973.

DURAN, MARCO ANTONIO, *La pobreza rural en una zona agraria crítica,* Centro Nacional de Productividad, México, 1971.

ECKSTEIN, SALOMON, *El marcomacroeconómico del problema agrario mexicano,* Centro de Investigaciones Agrarias, México, 1968.

FERNANDEZ Y FERNANDEZ, RAMON, *Cooperación agrícola y organización económica del ejido,* Secretaría de Educación Pública, Dirección General de Educación Audovisual y Divulgación, México, 1973.

————, *La reforma agraria y la ganadería,* Centro de Economía Agrícola, Chapingo, México, 1976.

————, *Renovación agraria,* Centro de Economía Agrícola, Chapingo, México, 1977.

————, *Temas agrarios,* Fondo de Cultura Económica, México, 1974.

FOOD AND AGRICULTURE ORGANIZATION OF THE UNITED NATIONS (FAO), *Agricultural Adjustment in Developed Countries,* Rome, 1972.

————, *Agricultural Commodity Projections, 1970–1980,* vols. 1 and 2, Rome, 1971.

————, *Agriculture Toward 2000,* Rome, 1979.

————, *Fertilizer Yearbook* (annually).

————, *International Agricultural Adjustment,* C73/15, Rome, 1973.

————, *L'Agriculture a temps partiel,* Paris, April 1971.

————, "Population, Food Supply and Agricultural Development," paper presented at World Population Conference, Bucharest, 1974.

————, *Production Yearbook,* Rome (various issues).

————, *Trade Yearbook,* Rome (various issues).

FRIEDMAN, SANTIAGO I., *Participatory Supporting Research for Production Increasing Programs in Mexico,* Rural Development Workshop, University of California, Berkeley, June 1975.

————, *The Role of Social Science Research in Rural Development: A View from Mexico,* Rockefeller Foundation Development, New York, April 1975.

GONZALEZ GALLARDO, ALFONSO, *El desarrollo del cultivo de la caña y la producción de azúcar en México en los 435 años de existencia (1538–1973) de la industria azucarera mexicana,* México, 1973.

GUANOS Y FERTILIZANTES DE MEXICO, S.A., *Estadísticas de ventas de productos por zonas* (various numbers).

GUERRERO, EUQUERIO, *Las contradicciones del amparo agrario,* Revista de Revistas, 19 May, 1976.

INSTITUTO DE INVESTIGACIONES SOCIALES (Autonomous National University of Mexico), *El perfil de México en 1980,* 2 vols., Siglo Veintiuno Editores, México, 1970.

INSTITUTO MEXICANO DE INVESTIGACIONES ECONOMICAS, *Los distritos de riego del noroeste, tenencia y aprovechamiento de la tierra,* Centro de Investigaciones Agrarias, México, 1957.

INSTITUTO NACIONAL DE INVESTIGACIONES AGRICOLAS (INIA). *Folletos técnicos* and *Folletos de divulgación* (various issues).

————, Centros de Investigaciones Agrícolas: del Bajío (CIAB), de la Mesa Central (CIAMEC), del Noreste (CIANE), del Noroeste (CIANO), de la Peninsula de Yucatán (CAIPY), de Sinaloa (CIAS), del Sureste (CIASE), de Tamaulipas (CIAT), de la Chontalpa (CIEACH); *informes, boletinos, folletos,* etc. of these research stations.

INSTITUTO NACIONAL DE LA NUTRICION, División de Nutrición, *Encuestas nutricionales en México* (various volumes and years).

————, *La crisis de alimentos en México,* J. Ramírez Hernández et al., México, January 1975.

————, *La estructura del consumo de alimentos en el medio rural pobre* (symposium by several authors), México, 1976.

ISZAEVICH, ABRAHAM, *Modernización de una comunidad oaxaqueña del valle,* Secretaría de Educación Pública, Dirección General de Educación Audovisual y Divulgación, México, 1973.

Ley Federal de Reforma Agraria, (edited by) Martha Chávez P. de Velázquez, Editorial Porrúa, México, 1971.

MENDIETA Y NUNEZ, LUCIO, *El problema agrario de México y la Ley Federal de Reforma Agraria,* Editorial Porrúa, México, 1975

NACIONAL FINANCIERA, S.A., *Statistics on the Mexican Economy* (various issues).

NAVARRETE, IFIGENIA M. DE (ed.), *Bienestar campesino y desarrollo económico,* Fondo de Cultura Económica, México, 1971.

ORIVE ALBA, ADOLFO, *La irrigación en México,* Editorial Grijalvo, México, 1970.

PALACIOS VELEZ, ENRIQUE, *Productividad, ingreso y eficiencia en el uso del agua en los distritos de riego en México,* Colegio de Postgraduados, Chapingo, México, 1975.

Problemas Agrícolas y Industriales de México, vol. 4, no. 3, México, 1952.

RAMIREZ FERNANDEZ, JUAN, et al., "Aspectos socio-económicos de los alimentos y la alimentación en México," *Comercio Exterior,* vol. 21, no. 8, August 1971.

————, Y CHAVEZ V., ADOLFO, "La disponibilidad de alimentos en México en el último cuarto de siglo," *Comercio Exterior,* vol. 18, no. 12, December 1968.

RODRIGUEZ CISNEROS, M., *Características de la agricultura mexicana,* Banco de México, 1974.

SANCHEZ FUENTES, MIGUEL, *Presente y futuro de la explotación cañera en México,* Centro Nacional de Productividad, México, 1976.

SCHUMACHER, E. F., *Small Is Beautiful,* Bland and Briggs, London, 1973.

SECRETARIA DE AGRICULTURA Y GANADERIA (see next entry).

SECRETARIA DE AGRICULTURA Y RECURSOS HIDRAULICOS, Centro Internacional de Agricultura Tropical, *Informe Anual* (various numbers).

————, *Informes de la Comisión Técnica Consultiva para la determinación regional de los coeficientes de agostadero* (one volume for each of the 32 states, from 1972 onwards).

————, Dirección General de Economá Agrícola, *Análisis general de las encuestas* (1976 and subsequent years).

————, ————, *Anuario estadístico de la población y producción pecuaria* (annually).

————, ————, *Anuario estadístico de la producción agrícola* (annually).

————, ————, *Consumos aparentes de productos agropecuarios, 1925–1976.*

————, ————, *El uso de los fertilizantes en los distritos de riego,* Informe Estadístico (annually).

————, ————, *Estadística agrícola en los distritos de riego,* informe Estadístico (annually).

————, ————, *Producción de la ganadería, de las industrias y de las semillas mejoradas en los distritos de riego,* Informe Estadístico (annually).

————, ————, *Sistema Alimentario Mexicano*, 1980.

————, ————, *Sistema general de encuestas*, 1976.

————, Dirección General de Extensión Agrícola, *Boletín Informativo* (various numbers).

————, ————, *El extensionismo agrícola en México, 1974*.

————, ————, *La extensión agrícola en México: qué es y qué hace*. Chapingo, 1974.

————, ————, *Memorias de los Congresos Nacionales de Extensión Agrícola* (various years from 1974).

SECRETARIA DE GOBERNACION, *La constitución política de los Estados Unidos Mexicanos, 1975*.

SECRETARIA DE PROGRAMACION Y PRESUPUESTO, Dirección General de Estadística (formerly attached to the Secretaría de Industria y Comercio), *Agenda Estadística* (annually).

————, ————, *Censo agrícola, ganadero y ejidal, resumen general* (for each of the years 1930, 1940, 1950, 1960 and 1970).

————, ————, *Censo general de población* (for each of the years 1930, 1940, 1950, 1960 and 1970).

————, ————, *Directorio de ejidos y de communidades agrarias (1970)*, México, December 1974.

————, ————, *Encuesta especial sobre rendimientos y precios medios*, México, 1972.

————, ————, *Resumen especial (ejidal)*, vols. 1 and 2, México, 1976.

————, ————, *Unidades de medida regional (del V censo)*, México, 1973.

SECRETARIA DE RECURSOS HIDRAULICOS (prior to its merger with Agriculture in 1977), *Recursos Hidráulicos* (annually; various issues).

————, Comisión del Rio Fuerte, Distrito de Riego No. 75, *Estadísticas de 1944–45 y adelante*, Los Mochis, Sinaloa.

————, Dirección General de Distritos de Riego, *Características de los distritos de riego*, vols. I, II and III (annually).

————, ————, *Costos de producción de los cultivos en los distritos de riego de la zona central*, Informe Estadístico, No. 65.

————, ————, *El uso de los fertilizantes en los distritos de riego* (annually).

————, ————, *Estadística agrícola del ciclo 1951–52*, Informe Estadístico (no. 7 and subsequent years).

————, ————, *La mecanización agrícola en los distritos de riego*, Informe Estadístico (annually).

————, ————, *Ley Federal de Aguas*, Publicación legal No. 13, México, 1972.

————, ————, *Salinidad de los suelos y calidad del agua de riego*, Memorandum Técnico No. 351. México, March 1973.

————, Subsecretaría de Planeación, *Plan Nacional Hidráulico 1975*.

————, ————, Documentación del Plan Nacional Hidráulico: *Diagnóstico de la migración en México;* also, *Regionalización e indicadores regionales*, México, 1976.

SOLIS, M. LEOPOLDO (ed.), *La economía mexicana*, Fondo de Cultura Económica, México, 1973.

URQUIDI, VICTOR L., "Empleo y explosión demográfica," *Demografía y Economía*, vol. 8, no. 2, El Colegio de México, 1974.

VENEZIAN, EDUARDO L. and GAMBLE, WILLIAM K., *The Agricultural Development of Mexico*, Praeger, New York, 1968.

WILKIE, RAYMOND, *San Miguel, a Mexican Collective Ejido,* Stanford University Press, Stanford, California, 1971.

YANES PEREZ, L., *Mecanización de la agricultura mexicana,* Instituto Mexicano de Investigaciones Económicas, México, 1957.

YATES, PAUL LAMARTINE, *Alternativas económicas y financieras,* Coloquio Internacional, Cocoyoc, Morelos, 1974.

——————, *El campo mexicano.* 2 vols. Ediciones el Caballito, S.A., México, 1978.

——————, *El desarrollo regional de México,* Banco de México, Departamento de Investigaciones Industriales, México, 1961.

YUKAWA, SETSUKO, *Transformación de la agricultura tradicional de México,* El Colegio de México, 1975.

INDEX